Autonomous Vehicles for Safer Driving

Other SAE books of interest:

V2V/V2I Communications for Improved Road Safety and Efficiency
By Ronald K. Jurgen
(Product Code: PT-154)

Automotive E/E Reliability
By John Day
(Product Code: T-126)

Automotive Software Engineering
By Joerg Schaeuffele and Thomas Zurawka
(Product Code: R-361)

Autonomous Vehicles for Safer Driving

By Ronald K. Jurgen

SAE International

Warrendale, Pennsylvania, USA

SAE International
400 Commonwealth Drive
Warrendale, PA 15096-0001 USA

E-mail: CustomerService@sae.org
Phone: 877-606-7323 (inside USA and Canada)
724-776-4970 (outside USA)
Fax: 724-776-0790

ISBN 978-0-7680-7993-7
Library of Congress Catalog Number 2013932495
SAE Order Number PT-158
DOI 10.4271/PT-158

To purchase bulk quantities, please contact
SAE Customer Service
e-mail: CustomerService@sae.org
phone: 877-606-7323 (inside USA and Canada)
724-776-4970 (outside USA)
fax: 724-776-0790

Visit the SAE Bookstore at
books.sae.org

Cover photo used with permission, courtesy of Toyota Motor Sales, U.S.A., Inc.

Dedication

This book is dedicated to my friend Richard Keaton.

Table of Contents

Introduction

Autonomous vehicles are no longer a distant future goal. Their future is now. This milestone has come about for a variety of reasons. Principal among them is the ability to incorporate numerous automotive technologies developed over many years, such as adaptive cruise control, forward collision warning, lane departure warning, and V2V (vehicle to vehicle) and V2I (vehicle to infrastructure) communications, into one complex system.

Reaching this goal has come about through impressive research, development, and testing procedures. In an achievement such as this, it is not surprising that it took cooperative ventures among car companies, universities, government agencies, and other entities in addition to single company efforts.

At first thought, however, one might be concerned about the safety of autonomous vehicles, but in truth they will be much safer than vehicles driven by humans. One obvious reason is that autonomous vehicles will not be distracted by such activities as texting on cell phones and other unwise activities that people engage in while driving, as well as their driving errors. In addition, autonomous vehicles can do what no human can do. Automatic interventions, such as those now routine with antilock braking systems and stability systems, will be built in to prevent accidents in case of a safety problem. For example, through the use of V2V and V2I technologies, autonomous vehicles can be warned and take action to avoid a traffic jam or accident around a bend and out of sight. And, finally, driver intervention will always be a choice if needed or desired.

Despite the many successes attained in these projects, major problems remain to be resolved. Keeping production costs of autonomous vehicles at a viable level is a major challenge. Municipalities worldwide need to legalize use of autonomous vehicles on their roads and highways. Perhaps most important of all, consumers have to be convinced that they want autonomous vehicles in their futures.

It is with all of these factors in mind that the papers in this book were carefully selected to bring the reader up to date on successful demonstrations of autonomous vehicles already accomplished, ongoing projects, and what the future may hold. To do so, the book is divided into three sections: overview, major design and test collaborations, and a sampling of autonomous vehicle research projects.

The comprehensive overview paper covers not only the current state of autonomous vehicle research and development, but also practical obstacles to be overcome and a possible roadmap for major new technology developments and collaborative relationships.

The section on major design and test collaborations covers Sartre, DARPA challenges, and the USDOT and the Crash Avoidance Metrics Partnership-Vehicle Safety Communications (CAMP-VSC2) Consortium.

The Sartre Project is a European collaborative project funded by the EC with participation of seven entities from four countries. It focuses on "design, set up, evaluation, and validation of road train systems that do not require modification to the actual road infrastructure, making use of available technologies for in-vehicle applied systems."

DARPA offered three challenges for unmanned and autonomous off-road ground vehicles that required self-funding. One of the papers presents Ohio State University's experiences at all three challenges. A second paper describes the development of a fully autonomous vehicle designed for urban environments developed and tested for under $20,000 in hardware costs. The final paper in this section describes the USDOT and the Crash Avoidance Metrics Partnership-Vehicle Safety Communications 2 (CAMP-VSC2) Consortium of Ford, GM, Honda, Mercedes, and Toyota. The consortium developed and tested vehicle safety systems using Dedicated Short Range Communications (DSRC) at 5.9 GHz in combination with vehicle positioning.

The final section presents seven papers detailing significant recent and ongoing research projects by individual companies on a variety of approaches to autonomous vehicles.

The authors of the overview paper conclude by stating "Are we now ready for a giant leap forward to the self-driving car with all its complexity and interdependencies? Humans will need to grow with and adapt to the technological advancements of the machine, and we'll deeply challenge our social and political paradigms before we're done. Even if we as engineers are ready, is the driving public ready?"

Ronald K. Jurgen, Editor

Overview

SAE International

Autonomous Driving - A Practical Roadmap

2010-01-2335
Published
10/19/2010

Jeffrey D. Rupp and Anthony G. King
Ford Motor Company

ABSTRACT

Successful demonstrations of fully autonomous vehicle operation in controlled situations are leading to increased research investment and activity. This has already resulted in significant advancements in the underlying technologies necessary to make it a practical reality someday. Not only are these idealized events sparking imaginations with the potential benefits for safety, convenience, fuel economy and emissions, they also embolden some to make somewhat surprising and sometimes astonishing projections for their appearance on public roads in the near future.

Are we now ready for a giant leap forward to the self-driving car with all its complexity and inter-dependencies? Humans will need to grow with and adapt to the technological advancements of the machine and we'll deeply challenge our social and political paradigms before we're done. Even if we as engineers are ready, is the driving public ready?

Putting a man on the moon was achieved through a series of logical extensions of what mankind knew, with necessity driving a search for technical solutions in the usual as well as unusual places, much as the Defense Advanced Research Projects Agency did with their Grand Challenges. This paper addresses the autonomous vehicle vision in terms of the current state and some of the practical obstacles to be overcome, and proposes a possible roadmap for the major technology developments, new collaborative relationships, and feature implementation progression for achieving those ambitions.

1.0. INTRODUCTION

The desire for the ultimate personalized, on-demand, door-to-door transportation may be motivated by improved personal convenience, emissions and fuel economy; yet there are also potential safety benefits from the pursuit of autonomous vehicles. This paper describes some of the practical obstacles in achieving those goals, and explores the use of near term applications of technologies that will be by-products of pursuing them. This includes a partial history of autonomous vehicle development (Section 2), potential consumer acceptability issues (Section 3), followed by a development roadmap and discussion of some variables to be addressed before autonomous vehicles become viable (Sections 4 and 5), and ends with a consideration of collaborative relationships that could assist in acceleration of development and issue resolution (Section 6).

2.0. THE CURRENT STATE - PUTTING THE HYPE INTO PERSPECTIVE

There has been escalating excitement about fully autonomous vehicles in the robotics community for some time and the excitement has now spilled over to the automotive industry. The idea of a self-driving, road-ready vehicle sparks the imagination, and is a familiar concept due to repeated exposures in popular culture; be it movies, cartoons, television, magazines, books or games.

An exhibit at the 1939 World's Fair in New York[1] presented a vision where cars would use "automatic radio control" to maintain safe distances, a depiction of transportation as it would be in 1960, then only 21 years into the future. One of the earliest attempts at developing an actual vehicle was led by Dr. Robert E. Fenton who joined the faculty at Ohio State University in 1960 and was elected to the National Academy of Engineering in 2003[2]. It is believed that his pioneering research and experimentation in automatic steering, lane changing, and car following resulted in the first demonstration of a vehicle that could drive itself. Since then,

OEMs, universities, and governmental agencies worldwide have engineered or sponsored autonomous vehicle projects with different operating concepts and varying degrees of success.

Most recently, the Defense Advanced Research Projects Agency (DARPA), an agency of the United States Department of Defense, sponsored three autonomous vehicle challenges. While a number of media friendly successes resulted in good 'photo ops', those in technical fields and many others readily appreciate the magnitude of work required to mature these vehicles into a viable, real world, design.

2.1. Contemporary Error Rates -- We're Way Off

In the months preceding the inaugural DARPA Grand Challenge in 2004, William "Red" Whittaker of Carnegie Mellon's Robotics Institute, with over 65 robots to his credit, stated "We don't have the Henry Ford, or the Model T, of robotics", "Robotics is not yet mainstream; it's not yet a national conversation."[3] His contributions and those of his students over the next few years would move the needle significantly, but his comments suggest the true nature of the challenge.

The error rates of robotically piloted vehicles today are still very high compared to human-piloted vehicles. At the 2005 DARPA Grand Challenge (DGC2) 5 of the 23 finalists successfully finished the 132 mile course, while two years later, at the 2007 DARPA Urban Challenge Event (UCE), 6 of the 11 finalists finished a 60 mile course. The mean mileage between significant errors (failure) at these events was 120 miles for DGC2 and 100 miles for UCE[4]. The errors cannot be attributed to a single primary cause, rather, multiple simultaneous causes and interactions including sensing, interpretation of the scene and simplification of its full complexity, simplifying assumptions and non-representative tradeoffs built into the algorithms, as well as unintended software bugs and hardware durability. Compare robotically piloted vehicle errors to that of human drivers, who averaged 500,000 miles driven between crashes in 2008[5].

Despite humans being 3-4 orders of magnitude better at driving than robots, crashes of varying severity occur regularly. In 2008 in the United States alone, there were 34,000 fatal crashes and 1.6 million injury crashes. Autonomous vehicles may need to be better drivers than humans, exhibiting fewer errors, to gain acceptance. The error rates inherent in today's autonomous vehicles are unacceptable for real world deployment in the present and will be for some time to come.

2.2. Progress Has Been Slow

Recalling the many predictions of a self-driving car over the last four decades, it is obvious that autonomous vehicles have taken and will take far longer than expected, especially when it comes to operational safety. Fully autonomous vehicles today are the product of laboratories, test tracks, and prize winning competitions, mainly conducted under favorable conditions with minimal and controlled uncertainties and no penalty for error. With limited success even in ideal situations, industry has little choice but to methodically split the problem into attainable steps, learning and developing the necessary enabling technologies along the way.

The combination of radio detection and ranging (RADAR) functionalities was patented by Christian Hülsmeyer in 1904[6], building on work from the mid-1800s by physicists James Maxwell and Heinrich Hertz. The majority of the development since then has been driven by maritime collision avoidance and military defense applications, including important signal processing extensions such as target velocity estimation based on frequency shift as proposed by physicist Christian Doppler. Despite this early start, it wasn't until 1999, with seven years of focused target tracking and controls development as well as electronics miniaturization, that Ford Motor Company launched the world's first-to-market radar-based ACC system with braking for an automotive application, on a Jaguar XKR.[7]

More than a decade later, advances in sensing technology critical for autonomous vehicle applications are just now accelerating significantly. Functionality of automotive forward-looking radars is increasing, even while prices are decreasing, with a drop of 75% over two generations expected in one case.[8] The progression to today's state of the art dual mode electronically scanned systems has allowed industry to use the resulting increased accuracy and availability to expand to new customer functions.

Digital camera systems have similarly been in existence for quite some time, with a patent application for "All Solid State Radiation Imagers" filed in 1968[9], and are now progressing more rapidly too. CMOS imagers have demonstrated increasing sensitivity, dynamic range, and pixel count, while costs have decreased due to the large volumes of consumer electronics applications. More recently, advancements in machine vision algorithms have enabled the evolution from lane tracking to significantly more complex vehicle and pedestrian detection and tracking functions.

Fusion sensing systems are also starting to see more automotive applications as well. Combining multiple sensing modalities, fusion leverages the orthogonality that can be established where the strength of one complements the weakness of another. This can create a sensing system with robustness and reliability greater than the sum of its parts. Ford developed and launched a radar-camera fusion system for Collision Avoidance Driver Support (CADS) functionality on the Volvo S80 in 2007. This was further expanded on the 2011 S60, overlaying a fused camera / forward looking multi-mode radar, with a multi-beam infrared and ultrasonic sensors, enabling collision warning and full auto braking for vehicles and pedestrians for collision avoidance, a world first, in addition to ACC, Lane Departure Warning, and Driver Alert (driver impairment monitoring) functionality.[10]

Other sensing technologies are also under development to better describe and interpret the external environment. Although automotive lidars, especially for ACC, have fallen out of favor, the development of 360° scanning and flash designs may bring about their resurgence. Detailed on-board maps are now available to help predict the road attributes ahead. Even as the number of radars and cameras in the vehicle proliferate, the industry also recognizes that on-board sensing could be significantly augmented through direct communication with other vehicles and the infrastructure. Research in the area of vehicle-to-vehicle and vehicle-to-infrastructure communications will be critical to any future cooperative transportation network. Despite these advancements, the verdict is still out as to the form of the ultimate sensing solution.

The majority of today's situation assessment algorithms enable only advisory and warning systems, as these systems are more easily implemented than fully autonomous control; using sensor data, the algorithms interpret the environment, predict the future, and provide some related driver support. With this limited approach, most performance errors merely result in annoyance. The environmental sensing system and control algorithm requirements are not as stringent as needed for autonomous operation, where the machine makes a decision and takes control of the vehicle. In the latter case, an incorrect decision may possibly result in a wrong action, possibly causing a collision when one may not have occurred otherwise. While designing a system that reacts positively (e.g. automatically applies the brakes prior to a collision) is readily achievable, the more difficult part of the task is to design the system to seldom make a mistake, and have the reliability and robustness necessary to appropriately respond to real world noise factors. The autonomous systems that exist today in controlled laboratories and test tracks are just not ready for the uncontrolled uncertainties of real world conditions. Automotive engineers are proceeding slowly to help ensure that appropriate level of performance exists before introduction.

2.3. Reluctant Consumer Acceptance of Autonomous Control

One need read only a few blogs in order to appreciate that consumers are uncomfortable with a machine making decisions for them and you can easily conclude that some drivers do not trust their vehicle taking even limited autonomous control. An independent analysis is available that describes the phenomenon of decision trust and the attributes affecting safety feature purchase.[11] Furthermore, the lack of third party endorsements for more than the most basic CADS functions (i.e. Forward Collision Warning; further enumerated in Section 4.2, Use Cases) has created little feedback for these technologies and therefore little customer enthusiasm and 'pull', and the lack of government mandates has created no 'push'.

Governmental and public domain agency action may help accelerate acceptance and adoption, or at least access and usage, of autonomous technologies, and several organizations around the world are considering regulation. Anti-lock braking systems were introduced in 1971, and reached 86% market penetration only after 37 years, in 2008. Compare that to Electronic Stability Control (ESC), introduced in 1995. Although the industry already had an implementation plan, the U.S. National Highway Traffic Safety Administration (NHTSA) accelerated penetration by mandating standard ESC in all new vehicles by 2012, less than 20 years later. NHTSA has included Forward Collision Warning and Lane Departure Warning in the ratings for the Active Safety New Car Assessment Program. The European Commission is considering mandates for Collision Mitigation Systems on light commercial vehicles. Non-governmental organizations such as the Insurance Institute for Highway Safety and the Consumers Union (publishers of Consumers Report magazine) have started to address CADS technologies, raising consumer awareness. Insurance companies are considering lower rates for vehicles with CADS features.

It is interesting to note that market adoption rates may have some cultural influence. Take the ACC system for example, a fairly straightforward extension of traditional cruise control that provides longitudinal control of the vehicle using brake and throttle to maintain distance to a vehicle in front. Ten years after initial introduction, it is finally getting significant mass market recognition, but the penetration rate in North America is only a fraction of that in Japan where the market seems to have a greater percentage of early adopters, allowing for rapid technology evolution. An independent study detailing these differences is also available.[12,13]

2.4. Today's Feature Implementation Progression

Although the adoption of CADS functions in private vehicles has been slow to date, the world is on the cusp of more widespread implementation of limited autonomous control. Technology will continue its rapid advance and as consumer acceptance expands, the industry will see systems that warn the driver of hazardous conditions, support driver actions, provide limited autonomous control with driver command, and even take some fully autonomous action to avoid a potential collision. The nature, direction, and pace of CADS feature introduction and progression can be inferred from the following list:

• Longitudinal support:

1958	Cruise Control (non-adaptive)
1971	Anti-lock Braking System (ABS)
1991	Ultrasonic Park Assist
1999	Adaptive Cruise Control (ACC)
2003	Forward Collision Warning (FCW)
2003	Collision Mitigation by Braking (CMbB),
2006	Stop & Go ACC (S&G)
2006	Full speed range ACC
2008	Low Speed CMbB (collision avoidance, City Safety™)
2010	Full Autobraking CMbB
2013 (*est.*)	Curve Overspeed Warning (electronic horizon-based)
2015 (*est.*)	Curve Overspeed Control (electronic horizon-based)

• Lateral support:

1971	ABS
1990	Variable steering assist, cross wind compensation, etc. (electrical)
1995	Electronic Stability Control
2001 (*Japan*)	Lane Departure Warning (LDW)
2001 (*Japan*)	Lane Keep Assist (LKA)
2002	Roll Stability Control (RSC)
2003 (*Japan*)	Lane Centering Aid (LCA)
2004 (*Japan*)	Intelligent Parking Assist System (IPAS)
2005	Blind Spot Information System (BLIS)
2006	Active Parking Assist
2007	Driver Alert, Driver Impairment Monitoring
2012 (*est.*)	Lane Change Merge Aid (LCMA)
2013 (*est.*)	Emergency Lane Assist (ELA)

• Integrated lateral and longitudinal support:

2010	Curvature Control (stability control-based)
2014 (*est.*)	Traffic Jam Assist (TJA) - S&G ACC + LCA

With the continuous evolution and improvement suggested by this feature progression, it is clear that many benefits from warnings and limited autonomous control are being realized, and more soon will be. Beyond this, incremental benefits can be reasonably attained only by advancing to a more complex and potentially intrusive level of functionality, one more closely associated with fully autonomous driver-support features. As suggested previously, consumer paradigms may need to shift again, and the governmental and social infrastructure may need to adapt. The key factor in establishing consumer comfort with these technologies may be empowerment of the driver in making the final control decision, say, overriding the function of the CADS feature.

3.0. A LOOK TOWARD THE FUTURE

3.1. Uncertainty, Unpredictability and Human Error

According to a World Heath Organization study from 2004, traffic accidents result in approximately 3,300 deaths every day, equaling over 1.2 million fatalities each year worldwide. By 2020, annual fatalities due to vehicular accidents are projected to increase to 2.34 million, assuming continuation of current trends. Already the leading cause of injury mortality, road crash injury is likely to become the third leading cause of disability-adjusted life years (DALYs) in the same time frame, trailing only heart disease and unipolar depression.[14]

The pursuit of autonomous vehicles, where drivers are supported in the driving decision making process, has a positive correlation with the pursuit of fatality-free, and even collision-free, transportation. Humans are fallible; driver error is the primary cause of about 90% of reported crashes involving passenger vehicles, trucks, and buses.[15] A misconception links these human errors solely as "... evidence of lack of skill, vigilance, or conscientiousness"[16] or insufficient training, since highly trained and skilled experts, such as doctors and pilots, are also susceptible to making errors, some with serious consequences. Frequently, errors result from poor reactions to unpredictable events and incomplete information as factors in the decision making processes. These probabilistic external factors typically form complex interactions creating random non-repeatable events. One study of airline pilots found that"... small random variations in the presence and timing of these factors substantially affect the probability of pilots making errors leading to an accident."[17]

Given these uncertainties, it seems unrealistic to assume that a decision making process, be it human or machine, will make the appropriate decision 100% of the time. Moreover, we must be cognizant of the fact that drivers are not machines and contemporary machines were shown previously to have not attained any where near the levels of holistic human cognition. Further, human reaction to the same exact external input will vary from individual to individual, and will therefore continue to be subject to unpredictable outcomes.

These external and internal uncertainties characterize the system inadequacies in which errors occurred, where the driver and the vehicle are only a portion of the overall transportation system. Rothe describes how the concept of a living system, one that adapts to change and achieves a new balance, can be applied to a driving scenario.[18] He suggests that an interactive relationship exists among the various system factors - biological (health/illness), psychological (doubt), social (seclusion), societal (norms), economic (lost wages), legal/political (arrest), other drivers and vehicles, the road infrastructure, and information regarding their status (weather and road conditions). Each of these factors set the stage for the other with recursive feedback between them. Focusing on a single factor merely distorts the situation without resolving it.

The implication from this is that a better understood and more tightly coordinated overall system will result in reduced levels of unexpected future events, and thereby a reduced likelihood of collisions. Nearly error free decision making is a very hard problem but it needs to be solved before an autonomous vehicle system that provides 'Full Driver Assist' is ready. Predicting when it will be feasible is merely guess work, but a roadmap would still be useful in approaching it in a comprehensive and systematic fashion.

3.2. Autonomy in Other Transportation Modes

The Shinkansen railway system in Japan provides an example of a positive attempt and outcome. Running on separate track from conventional rail, the lines are built without crossings, use long rails that are continuously welded or joined with expansion joints that minimize gaps due to thermal conditions, employ Automatic Train Control for on-board signaling, have early warning earthquake detection so trains can safely stop, and enforce strict regulation with stiff fines to prevent trespassing on the tracks. From the train sets, to the tracks, the operators, the information availability, and the governmental regulations, this tightly controlled system is designed to reduce the amount of uncertainty and enable a high reliability of safe decision making. The result: no injuries or fatalities due to derailment or collision in 46 years of operation, and only one derailment (with no injury) caused by an earthquake in 2004, while carrying over 150 million passengers a year (in 2008).[19]

The Shinkansen system demonstrates that fatalities may not be an inevitable consequence of transportation after all. A major difference lies in the train operators themselves - besides being highly trained, their number is but a mere fraction of the billions of personal-vehicle drivers in the world today. Thus, tight control over the system includes control over this uncertainty: the variance of individual driver (operator) reactions to external inputs. In the quest for further reductions in collisions in private vehicles it is inevitable to eventually seek to replace human unpredictability with something a bit more predictable. The result may not be purely an electronic substitution, but rather a driver 'subsystem' that involves both the human and the electronic system. The electronic system informs and aids the human in the ways it is better suited, by leveraging its strength (e.g. estimating range and closing velocity), and leaving higher level tasks for the human 'driver' to perform. It's an orthogonal decision making mode, similar to fusion of multiple modalities of sensing (e.g. radar and vision). Each has its strengths and weaknesses, but when properly combined results in a more reliable and robust solution.

Consider another self-driving (autonomous) vehicle, one that has existed for centuries. A ship's captain is on board, but may never touch the wheel; he is in command but not necessarily in direct control. He has a surrogate system, in this case human, that is 'programmed' to carry out 'lower level' control functions, whether that human be a helmsman, quartermaster, or engine room operator, relieving the captain of the burden of continuous interaction. Similarly, you hire and 'command' a taxi as a system (car + driver) by requesting a destination, but there is no direct control.

Beyond those analogies, there are many 'self driving vehicle' applications in existence today. These are autonomous vehicles in a very real sense, some having greater autonomy than others. Commercial airplane pilots engage the autopilot and monitor the systems until direct intervention is needed, whether induced by tower commands or an emergency. Automated train systems, such as those within an airport terminal network, ferry people without an onboard pilot, but are still controlled by humans at a central station. The military has significant autonomous vehicle assets in operation today including Unmanned Air and Ground Vehicles (UAV and UGV respectively), which can be directly controlled by a remote operator, or programmed for autonomous operation to patrol a certain area for a set number of hours, for example. In the past few decades, modern sailing vessels have replaced mundane operator tasks with computer control; an autopilot can now navigate from waypoint to waypoint and seamlessly control throttle, rudder, and roll stabilizers.

All these are examples of vehicles with autonomous control, but still not completely without human oversight. If a pilot is not directly on board, then there is an operator monitoring remotely. There is no vehicle or transportation or mobility system that doesn't have human oversight of some sort. And we should expect the human operator to be 'in-the-loop' for a very long time.

Moreover, these semi-autonomous systems rely on operators trained specifically for driving. To become a commercial airline pilot, for example, one must first obtain a commercial pilot license after 250 hours of flight time, with allocations dedicated to specific conditions and maneuvers. Additionally, a commercial pilot needs an up-to-date first- or second-class medical certificate, an instrument rating and a multi-engine rating. Thousands of additional flight hours are needed to even be considered for hire at a commercial airline. Once hired, additional training begins. Typically a 10 week course ensues, followed by a few weeks in the simulator, where the trainee experiences just about every emergency and anomaly imaginable. Once this training is done, initial operating experience is gained by flying some 25 hours with a special instructor pilot, followed by another flight test. Now the pilot can become a crew member. In order to become the captain of a major commercial airliner, a pilot must then obtain an airline transport pilot certificate which requires passing a written test, and logging 1,500 flight hours including 250 hours as the pilot in command. Similar levels of training are required to pilot a ship, control military UAVs, or control NASA's unmanned vehicles. Current driver training for operating an automobile is not nearly so stringent.

3.3. Do We Want a Driverless Car?

When people talk about fully autonomous vehicles, a common image is that of a driverless car, like the autonomous trains in an airport or DARPA challenge robots. Do consumers want a car without a driver, a car that can go somewhere without you like a military mission, whether delivering a package or picking up the kids after school with no one in control on board? There may be a few cases where a consumer wants someone or something else to do these tasks, but we already have services in place for that - package delivery services, buses, carpools, taxis, etc.

Instead, consumers of private autonomous vehicles may not want a driverless car, but rather a car that drives itself. You are in the car, and the car transports you, your belongings and your family, but you don't necessarily want to directly pilot it. You don't want to be locked into the detailed, sometimes tedious, moment to moment tasks of driving, but instead merely want and need to direct where it goes and how it gets there. You want command, but are willing to relinquish the detailed control to automation so that you can do something else; listening to music, placing a phone call, watching a movie, or just enjoying the scenery. What the consumer really wants is not a self-driving car, but an autonomous vehicle system that provides Full Driver Assist.

Recall the first mainframe computer, first PCs, the first PDA, and then the first cell phones. These devices were going to make our jobs and lives easier. And they have - not by doing work for us as originally thought, but by helping us work more efficiently. At first they were all clumsy devices, difficult to use, and not well accepted. Eventually they are integrated into the connected world in which we live and evolved into productivity tools that enable us to work in more places, more of the time. Similarly, the advent of autonomous technologies in vehicles will result in drivers that are more engaged in some aspects of the driving process rather than further removed, providing them with greater capability in managing the overall process. The driver would now be much more akin to the captain of the ship, biased toward the tactical, strategic, interactive, and predictive roles while leaving the role of the helmsman, lookout, navigator, and even quartermaster to the vehicle systems. Handling this type of automation in everyday life, however, requires that the consumer paradigm change.

3.3.1. Driving to a Seamless Experience

Smartphone owners can buy a special application (app) for just about anything, from checking the weather to checking your bank accounts and paying bills, from playing games to updating your social network and checking sports scores, and so on. There are dozens of apps just for social networking - one for each online site - plus apps for email, contacts, text messaging, and instant messaging. In today's smartphone implementation, the entire task of staying in touch with a social network is an exercise in opening and closing apps, which is a clumsy and overly complicated interface at best. Soon there will be a single app where you can see all your friend's updates on the social networking sites, while tying it seamlessly together with the contacts, photos, email, and text messages on your smartphone.

Like consumer electronics, the automotive industry is now tackling these issues; focusing on improving the in-vehicle experience by combining these apps into seamless experiences. MyFord Touch™, Ford's new driver-connectivity technology, complementing SYNC®, Ford's device and off-board service connectivity technology, is an example of integrating and simplifying the experience of entertainment and connectivity in the vehicle. Through the digital cluster displays, large touch-screen interface and voice interaction, the system allows the driver to naturally command the vehicle to play new music, seek traffic, direction and journey-related information, answer calls, make calls, and even listen to text messages through multi-modal interfaces. Software application programming interfaces (APIs) will soon be available to allow apps like Pandora and Stitcher to be controlled through the voice-controlled SYNC® system to stream audio to build a consistent, user-friendly interface within the vehicle itself.

This development progression repeats a trend that has occurred time and time again. Compare these steps for starting a Ford Model T[20] with today's 'turn the key' or 'push the button' ignitions:

1. Pull the choke adjacent to the right fender while engaging the crank lever under the radiator at the front of the car, slowly turning it a quarter-turn clockwise to prime the carburetor with fuel.

2. Get into the car. Insert the ignition key, turning the setting to either magneto or battery. Adjust the timing stalk upward to retard the timing, move the throttle stalk downward slightly for an idle setting, and pull back on the hand brake, which also places the car in neutral.

3. Return to the front of the car. Use your left hand to crank the lever (if the engine backfires and the lever swings counterclockwise, the left arm is less likely to be broken). Give it a vigorous half-crank, and the engine should start.

Development focuses on the task the consumer is trying to perform, and works to improve the overall user experience associated with that task. Through integration, the functional evolution simplifies the operation and significantly enhances the efficiency in performing that task. Historically, the movement towards a simplified, seamless experience to improve operating efficiency has been a key to widespread adoption of new technology, stimulating a series of consumer paradigm shifts. Similar to a smart phone, the technologies discussed in Section 2.4, Today's Feature Implementation Progression, may be considered standalone apps as well, but in a vehicle environment. Many of the highest technology features have had limited take rates possibly due to perception of cost, complexity and uncertainty of performance, but we expect this will benefit from development into a more seamless experience. Traffic Jam Assist is a technology that operates the distance control of ACC S&G in conjunction with the lateral control of LCA at low speeds. A later step will be to integrate all CADS functions into a comprehensive Full Driver Assist functionality, simplifying the web of complex CADS functions into a coordinated holistic system - user-friendly, easy to understand, and available to all consumers.

When done well, this advanced development can result in recommendations by opinion leaders at many levels, improving the familiarity and comfort level with the technology, further speeding adoption and penetration into everyday life. But what does Full Driver Assist really mean to consumers? What tasks do automotive consumers wish were more efficient?

3.3.2. Of Desires, Expectations, and Values

America has always been a country where motoring nostalgia is heavily intertwined with the freedom of exploration. This explains American's love affair with the car; with hands on the steering wheel, foot on the accelerator, and hair blowing in the breeze while cruising down Route 66. Americans are in their cars a lot - an average of 87 minutes per day according to an ABC News survey.[21] Some automakers have recently focused on remaking car interiors like a comfortable and luxurious living room, but driving is not all for fun.[22] Commuting to and from work comprises over 27% of vehicle miles traveled, more than any other category. The next highest category was social/recreational travel, including going to the gym, vacations, movies or theater, parks and museums, and visiting friends or relatives; i.e. using the vehicle as a means to get to a destination. These two categories alone comprise over 50% of all vehicle miles traveled. A recent study by Northeastern University indicated that, given past history, one can predict anyone's travel route and location with 93% accuracy.[23] These studies imply that people are repeatedly visiting, or commuting to, the same locales with significant regularity.

So do people enjoy the daily driving routine? The study by ABC News indicates that nearly 60% of people like their commute, but only if the trip is relatively easy. Nearly 4 out of 10 state the primary reason they like their commute is that it gives them quiet or alone time, and nearly a quarter identified that their commute is easy and has little congestion or traffic. For city dwellers with more than a 30-minute commute or experience traffic congestion, the percentage who likes their commute drops into the 40's. To further understand consumer behavior, it's necessary to understand the human emotion and values. A great majority of drivers, according to this study, at least occasionally feel very negative emotions while driving, with 62% feeling frustrated, 56% feeling nervous about safety, and 43% even feeling angry. But the same survey also says that 74% often feel independent, while 48% often feel relaxed while driving. Interestingly, independent and relaxed are not really emotions, but relate to core human values. The Rokeach Value Survey (RVS) identifies 18 terminal values, which are values every human strives to experience at least once in their life (and more often if possible), and 18 instrumental values, which are the preferred means of achieving those terminal values.[24] Independence is an instrumental value, and relaxed can correlate to inner harmony, a world at peace, or comfortable life terminal values.

These values seem to at least partially explain, if not directly motivate, people's desire to drive. They explain the high consumer demand for infotainment in the car -- drivers want to enhance relaxation through music or conversation. Infotainment systems, as a relaxing agent, will become even more important as traffic congestion worsens. Hours spent in traffic delays have increased 50% from the last decade and continue to increase[25], so it is expected that the number of people feeling relaxed while driving might actually decrease, even with infotainment systems in the vehicle. On the other hand, Ford and MIT's AgeLab, in conjunction with the U.S. Department of Transportation's New England University Transportation Center, have been working since 2004 to develop vehicle systems that detect the stress level of the driver at key points in time.[26] A recent extension of that project intended to identify specific stress-inducing driving situations, apply biometrics to monitor driver reactions and evaluate methods to incorporate new stress-reducing or even stress-optimizing features.[27] These features include the Blind Spot Information System with Cross Traffic Alert, Adaptive Cruise Control and Collision Warning with Brake Support, MyKey, Voice-Activated Navigation, and SYNC®.

Additionally, the RVS values discussed previously explain why only 5% of trips are on public transportation. Although one can just as easily feel relaxed on a commuter train as in a vehicle, 93% find traveling by car more convenient. It is this convenience that keeps drawing drivers back to the road; the freedom to leave whenever you want; the convenience of getting you from exactly point A to point B without changing modes of transportation. Having your own personal vehicle translates to independence, eliminating the need to rely on someone else to accomplish your own tasks or pursue your goals.

What do drivers want? They want a utilitarian appliance that moves them from door-to-door on their terms; they want to be more effective in the driving process, and they want luxury comforts. They use descriptors such as 'productive', 'efficient', 'relaxing' and 'personalized'. An autonomous transportation device with independent supervisory control would fit the bill, but they also want the ability to drive the enjoyable drives which may add excitement and enhance a sense of freedom. A successful vehicle will likely need to seamlessly blend full assist and fully manual modes of operation and probably everything in between to satisfy consumer needs, expectations, desires, and values.

3.3.3. Consumer Paradigms

In order to build the future of personal transportation that people want, the associated consumer paradigms must change. There is precedence for the shift necessary for adoption of new technological innovations. When Nicolas Joseph Cugnot introduced one of the first self-powered vehicles in 1769 (which was commissioned by the French army), not many imagined that this curiosity would spawn a technological gold rush for the next century and a half in a race to provide 'auto-mobile' vehicles to the masses. Instead there were concerns about their safety and usefulness, as this early vehicle could only travel at 2.5 mph for 10 minutes at a time, and crashed in its first demonstration. Technology progressed, and by the first half of the 1800s there existed a small market for steam-powered auto-mobile vehicles. However, in 1861, the British Parliament was sufficiently concerned about public safety to enact The Locomotive Act that severely limited operation of motorized vehicles on-road. Although this stopped most motorized vehicle development in Britain, innovation continued elsewhere, especially in Germany, France and the United States. As the automobile moved into the mainstream and garnered ever more press coverage, consumers became more comfortable with and confident in the technology. This Act was partially repealed in 1896, and automobile development accelerated at the turn of the century with the advent of electric and internal combustion propulsion. By 1913, Henry Ford was building Model T's that every working man could afford, the result of standardized manufacturing and internal combustion engine technology.

The evolution from the driver-guided to the autonomous personal vehicle will parallel the evolution from the horse-drawn to the auto-mobile carriage: a period of initial caution and low acceptance, initial innovation and invention, use by early adopters, followed finally by rapid innovation and expansion, mass market penetration, and standardization. New technology will deeply challenge the social and political paradigms of the day, but now, as always, humans will adapt. As before, full consumer acceptance will not occur until consumers observe early adopters for a sufficient amount of time to trust that the system can operate safely and has a mature level of robustness and functional tuning. The wall of resistance to limited autonomous control is just starting to fall. With consumers showing signs of increasing comfort with automation, expect acceleration in the implementation and penetration of vehicle CADS technologies. Each generation of CADS implementation builds consumer confidence in the technology, and eventually consumers will accept autonomous control as naturally as they accept a self-powered (auto-mobile) vehicle.

4.0. DEFINITION AND ROADMAP FOR A FULLY AUTONOMOUS VEHICLE

Successful development of something as complex as a fully autonomous vehicle will be most readily achieved by those taking careful evolutionary steps, rather than one revolutionary leap. The DARPA Challenges served to jump start work on autonomous vehicles in the commercial sector, and fed new learning back to the military-industrial complex that has been working on the same problem for decades. These competitions and demonstrations provide glamour and some important lessons, but the technologies developed will not be directly applicable to the consumer market for quite some time, if ever. They just are not the practical next steps to putting something into production for public sale; these solutions leap right past more fundamental problems.

However, there's a place for the revolutionary vision, partly to show the world the march towards autonomous control, but mostly to motivate the effort and the long-term investment required. Industry and society both need high visibility demonstrations to sustain enthusiasm through the arduous hours of detailed engineering and analysis necessary to turn a dream into reality. We need to take time to understand true consumer values, and then engineer the technology and infrastructure for the reliability and robustness necessary to enact a safe and secure driving experience, one that inspires consumer confidence.

An on-demand, door-to-door, personalized automated transportation system may very well be achieved some day, but there are many lesser autonomous functionalities that customers will value that can be implemented much more quickly. As the industry researches and engineers towards Full Driver Assist it needs to follow a spiral development model, spinning off technologies and capabilities as they mature, bringing the consumer along step-by-step, little by little. These spin-offs cannot be limited to only the latest and greatest technology implementations. They must also include low cost solutions that can be implemented on lower cost vehicles for global implementation.

What follows is one promising roadmap for realizing a fully autonomous vehicle, or more precisely a Full Driver Assist-capable vehicle. It begins with an overarching design philosophy followed by customer-valued Use Cases that build upon existing collision avoidance and driver support features, which should be sequentially achieved, with appropriate operational reliability and robustness before proceeding to successive levels.

4.1. Design Philosophy

Until we have proven sufficiently reliable machine automation in a highly complex, continuously varying, unpredictable environment, one filled with both human and autonomous agents, the approach should be to keep the driver in the loop, as well as in the driver's seat. The driver should have the responsibility to engage the Full Driver Assist feature in a manner similar to how Adaptive Cruise Control (ACC) is currently engaged; by selecting certain operating parameters such as headway and vehicle speed.

During hand-off transitions, the driver will be expected to maintain vigilance and readiness to take control of the vehicle and will need to be supported in doing so. To accomplish this, the Human Machine Interface (HMI) must evolve from the current set of least/latest credible/imminent hazard warnings intended to minimize nuisance alarms, to providing more immersive situational awareness throughout the driving experience. Experience with automated aircraft cockpits[28] reveals that operators are often uncertain about its 'behavior'. What is it doing now? What will it do next? How did I get into this mode? I know there is a way to get it to do what I want, but how? The potential for automation success increases when several situations are created:

• Timely, specific feedback is given about the activities and future behavior of the agent relative to the state of the world,

• The user has a thorough mental model of how their machine partner works in different situations,

• Automated systems take action and act consistently with prior direction from the human operator.

The driver has legal responsibility for control of the vehicle and must have the ability to override the system by adding or subtracting steering input, applying the brake or adding throttle. He will have the ability to request or make certain maneuvers (e.g. initiate a lane change), and may be requested to confirm appropriateness and acceptance of a system recommended maneuver.

4.2. Use Cases

Although potentially interpreted as a simple roadmap or a checklist of sequential developments, each step may very well require extraordinary advancement in order to attain the necessary operational reliability and robustness in increasingly complex operating scenarios. As discussed in Section 2.1, Contemporary Error Rates - We're Way Off, autonomous vehicles will likely need to be better drivers than humans, exhibiting even fewer errors and more favorable error modes before they gain initial acceptance, let alone widespread implementation.

Use Case 0.0 - Status Quo

This case exists in the majority of vehicles on the road today. There are no on-board radars or cameras to measure the external environment, and no algorithms to provide information, advice, warning, or control.

In this case, the vehicle operator is left to his own preferred behaviors, behaviors that can change from day to day or moment to moment based on many and various external and internal factors, varying from relaxed to assertive and even unaware driving. Opportunities exist to provide timely advice or assistance to the driver in making the most appropriate decision in the given situation. Such decision making would require vehicle systems that are equipped with algorithms that can learn from the past driver's experience, identify hazard situations, and accordingly implement the corresponding emergency maneuvers.[29] We can expect more on-board algorithms for driver and situation learning, anomaly detection, probabilistic decision making, and more intensive interaction between the driver and the electronic vehicle control systems in the future, resulting in an increased level of intelligence of the electronic vehicle control systems.[30,31]

The addition of external environment-sensing capabilities to vehicles enables the following use cases:

Use Case 1.0 - Information, Advisory and Warning

This set of use cases comprises advisory and warning CADS functions that help the driver make better decisions. The CADS function provides information and advisories to the driver about the road environment as well as warnings about potentially hazardous conditions, such as the possibility of an impending collision, without any autonomous vehicle control actions being taken.

Use Case 1.1

In this use case, the CADS functions address the road environment. The information is not critical to the driving task, but will help the driver make informed decisions in the near future. These advisory functions could include speed limits, sharp curve ahead, blind spot information, ultrasonic park aid, etc.

Use Case 1.2

In this use case, the CADS functions address potentially hazardous conditions, such as the possibility of an impending collision or low mu conditions ahead. These warning functions include Forward Collision Warning, Lane Departure Warning, Lane Change Merge Aid, etc.

Use Case 2.0 - Emergency Control

This set of use cases comprises autonomous emergency countermeasures that help the driver mitigate or avoid a potential collision. It is useful to separate autonomous emergency action from normal steady-state vehicle control because the control logic tends to be considerably different. Whereas emergency action is taken with the focus on collision avoidance, normal driving focuses more on passenger comfort and smoothness. This emergency action is only taken when there is an error in the normal driving state, whether internally or externally imposed; an autonomous emergency action could be taken, regardless of whether the car is under driver control or fully-automated control. Many functions that are a part of this use case have been deployed in vehicles around the world, albeit at fairly low take rates.

Use Case 2.1

In this use case, the CADS functions support driver actions to avoid a potential collision. These functions include brake assist, brake pre-charge, and limited autonomous braking to reduce the collision speed.

Use Case 2.2

In this use case, the CADS functions autonomously take corrective action to avoid an otherwise unavoidable collision, only acting at the last possible moment. These autonomous collision avoidance functions include ESC, RSC, LKA, and autonomous braking such as that introduced on Volvo vehicles as City Safety™ (launched in CY2008) and Collision Warning with Full Auto-Brake (with up to 25kph speed reduction, launched in CY2010).

Use Case 3.0 - Steady State Control

This set of use cases comprises the first stage of Full Driver Assist in normal steady state driving. CADS functions in this family comprise limited autonomous control for a short interval at the driver's command, allowing the driver to focus on other aspects of driving. These functions are designed typically for a specific driving scenario, and the driver will need to take over once the expected scenario is compromised.

Use Case 3.1

In this use case, the CADS functions take limited autonomous control in a single axis when activated by the driver. Functions in this use case, many of which are in production today, include ACC (longitudinal control, freeway driving), LCA (lateral control, freeway driving), S&G (longitudinal control, traffic queue), etc.

Use Case 3.2

In this use case, the CADS functions take limited autonomous control in multiple control axes when activated by the driver. Functions in this use case include Traffic Jam Assist (a pre-emptive assistance during traffic jams, i.e. S&G ACC plus low-speed LCA), combined with autonomous driving from expressway entrance ramp to exit ramp, where the driver gets onto the freeway and enables the system to drive to, but not exit at, the desired ramp.

Even this use case can have phased introduction, starting with short intervals, i.e. 'take the wheel' until circumstances change appreciably. This would be 'on demand' by the driver, but with system concurrence that would take into account traffic density and road geometry, with the vehicle driving in automatic mode at posted speeds without lane changes.

The short interval can be extended further to full entrance-to-exit ramp driving, lane changes and even passing, but which might be limited to roadways that the vehicle has already successfully driven passively and analyzed as 'self-drivable' to verify road markings, GPS availability, number of lanes, etc. The system may still ask the driver for confirmation, possibly having started a conversation with the driver via SYNC®, "Of the standard options (provide list) which would you like?", and extend to "I recommend changing lanes, shall I go ahead and do that for you?" or "Do you concur that it's ok to change lanes now?"

Additional extensions of this use case can include auto-park, latch, and platooning functionality.

Autopark is where the driver and passenger depart the vehicle and engage an autonomous valet parking routine in a known infrastructure space with administratively restricted access for pedestrians, etc. Latch is where a vehicle strictly follows a selected forward vehicle at a standard following distance, initially at a low speed (e.g. TJA), then gradually at higher speeds. Platooning, the automatic following of a 'certified' lead vehicle, such as a commercial bus or truck, is further enabled by V2V communication with and between the lead and following vehicles, characterized by latch functionality and close quarters/shortened following distance for fuel economy benefits.

Use Case 4.0 - Transitional Control

This use case is highlighted by new functionality that helps the driver negotiate challenging traffic. This includes scenarios where vehicles come together in potentially conflicting intent and space. Support is provided either through information, advice, warning, or automatic control, both as late evasive actions as well as early smooth coordination and cooperation.

Use Cases 4.1 and 4.2 - Freeway and Intersection Blending

The first case aides the vehicle activity at a freeway on ramp and off ramp, extending the steady state control from freeway ramp-to-ramp to include merging and exiting. This includes anticipation of the exit and the pre-positioning of the vehicle in the appropriate lane, i.e. actively pursuing a lane change, as opposed to passively recognizing a lane change opportunity. This also includes a second case for turning and merging into similarly flowing traffic at an intersection.

Use Case 4.3

This use case is characterized by aiding the driver when traversing intersections with opposing flow traffic. The functions will inform, guide, or even control by assessing whether crossing traffic will collide, pass in front, or pass behind; thus determining the safe margin for a left turn across oncoming (head-on) traffic as well as the safe margin for entering into traffic from a branch intersection, such as turning left across oncoming traffic from the left or simultaneously merging with oncoming traffic from the right.

Use Case 4.4

This use case addresses convenience support at an intersection. More specifically, this includes the automated slowing and stopping for a stop sign, yield sign, traffic light, prioritized junction (e.g. driveway connection with roadway), or other traffic management system or protocol in a preplanned comfortable fashion when there is no preceding traffic that would otherwise govern free flow. This is in contrast to emergency-based intersection transition functionality.

Use Case 4.5

In simple terms, this use case involves the 'safe stop', appropriate as a bootstrap function in the event the driver becomes totally disengaged, unresponsive, or incapacitated with respect to performing further driving tasks. This function communicates an emergency situation to surrounding traffic followed by the slowing, stopping, and parking of the vehicle on the side of the road. This is a marginally preferred alternative to continuing non-stop without driver intervention or stopping in-lane.

Use Case 5.0 - Revisiting Known Destinations and Routes

This use case is highlighted by the extension to all roads, no longer biased to limited-access expressways. However it is still restricted to roadways that the vehicle has already visited and passively assessed; where the vehicle is familiar with these surroundings and only has to confirm, rather than recognize and analyze, the proper way to interact with this new environment.

Use Case 5.1

This use case is limited to areas frequently traveled, for example from home garage to work parking lot, and therefore has high confidence in familiarity and low likelihood of change in the nature and condition of the infrastructure, accompanying traffic flow, etc.

Use Case 5.2

The next increment could be related to a vacation or holiday destination, say a weekend or summer cottage or condominium; a place it has already been but with longer distances and less frequently visited, introducing the greater possibility of changes since the last time it drove there. The ability to recognize changes in infrastructure and nature of traffic flow is correspondingly increased.

Use case 5.3

A special use case would be the local shuttle scenario. The uniquely tailored character of this scenario would provide the first opportunity for full drive-for-me functionality. This use case would be a limited pre-implementation feasibility demonstration and learning opportunity only, where the new HMI and situational awareness and autonomous controls can be further developed for reliability and robustness. Besides the driver being on board, there would also be a specially trained test co-pilot who is there only to intervene on the driver's behalf if warranted. The driver would be observed for tendency toward non-driving activities given this level of driving support and HMI. If the vehicle runs into a scenario it hasn't encountered before, or has not been designed to handle, or when sensing becomes blocked and the vehicle goes into 'limp home' mode, the driver can take over and continue the shuttle delivery manually, etc.

A shuttle such as this could be administratively managed by and wholly contained on a private road network, such as at the Ford Research & Engineering Center in Dearborn, Michigan. In this case it could build on the current Smart Intersection,[32] which would allow for greater adaptation of the vehicle and infrastructure for experimentation in terms of infrastructure communication, dedicated localization targets at road edges and intersections, etc.

Use Case X.0 - Traversing Unknown Routes and the General Case

Here is where we put it all together, pursuing the idealistic fully autonomous functionality. Autonomous, Full Driver Assist functionality is extended to situations that have not been sensed, analyzed, or hardcoded previously. The vehicle is capable of traveling anywhere; to places it has never been before, handling scenarios never encountered before -- it's ready for the all new experience.

In order to proceed to this level, the engineering staff will have learned through all preceding technology development cycles and use cases. The sensing hardware/software, as well as assessment software, will have been shown to be reliable and robust in the prior use cases, and are now stretched to modes where safe, real time learning is permitted, enabled, and successfully achieved using advanced machine learning algorithms. Fully autonomous functionality should achieve at least the same outcome as the human driver when encountering new situations, but with the greater diligence and situational awareness, as well as rapid recognition of subtle novelty that a machine can have.

Learning safely will depend on continuing development of HMI concepts through successive use cases. Cases that now merely communicate unlearned situations to the driver will be continuously succeeded by more complex, autonomous designs that further offload the driving task as a design ideal. The focus will be on the development of models and algorithms that are not only able to learn but also to summarize identified relationships and facts to a higher level of abstraction. The goal is to integrate this part of the multi-attribute decision-making mechanism under different conditions and situations which is a necessary condition for autonomous driving.

As previously discussed and shown in the market, CADS warning and emergency functions have been introduced in phases of gradually increasing effectiveness:

• CADS 1 - capability sufficient to warn only for moving cars/trucks/motorcycles,

• CADS 2 - capability to warn and provide relatively small autonomous braking action for stationary, as well as moving cars/trucks/motorcycles,

• CADS 2.1 - capability for large autonomous braking in reaction to vehicles ahead (special low speed case),

• CADS 2.2 - capability to both warn and initiate a large autonomous braking action when an alternative steering path is not available,

• CADS 2.3 - warning capability for unintended lane departure or potential impairment based on the driver's lateral control performance, and

• CADS3 - capability to both warn and initiate a large action in reaction to both moving and stationary cars/trucks/motorcycles and pedestrians.

In this use case, we build upon the level of effectiveness of the already available CADS functions and incremental use cases listed previously, and now extend them to the general case. The general case includes warnings and large autonomous actions (longitudinal and lateral) for hazards of all types including trees, poles, and other undefined or unexpected (e.g. debris in the driving lane) hazards, not just a smaller set of pre-classified types. The goal is to do this with early recognition and small actions for a smooth, seamless experience, vs. a panicked, last moment, large evasive emergency maneuver.

Intersection traversibility and cooperation, initially limited to conventional 3 or 4-way orthogonal configurations, is now extended to the n-way configuration. Scenarios may develop in such a way that the vehicle cannot brake to avoid a stopped car or large animal entering the lane, requiring an assessment whether it is safe to change lanes, e.g. whether there is parallel or oncoming traffic. Assessment of a 'safe alternative path' that may not be the designated driving surface, but which is suitable in emergency situations, such as the road shoulder, is also added. Implied in earlier use cases is the notion that late warnings of impending undesirable situations (a 'stop, don't do that' warning), will gradually be replaced with earlier advice, followed by increasingly stronger recommendations and requests for a positive desirable alternative action ('do this instead'), providing specifics the driver should focus on.

The CADS functions are also extended to the general case, including the full variety of weather and road conditions. Extreme weather conditions include snow where boundaries between driving and adjacent oncoming and non-driving surfaces are completely obscured. Road conditions include rural roads with painted lane markings only on the centerline, markings that may be faded, sporadic, or nonexistent, and gravel roads where the lane and road edge has no geometrically defined transitions whatsoever. Other extremes include off-road trails, stream fording, and open-spaces such as countryside, dunes, desert, tundra, etc.

5.0. SOME CONSIDERATIONS FOR BUILDING THE SYSTEM

Creating a system for autonomous personalized transportation involves more than just replacing one sub-system with another, replacing a driver function with an automated one, or completely replacing the human driver with a computer, let alone a robot. It will involve creating new subsystems, as well as new ways of integrating them; sub-systems that deal with interpretation of complex and cluttered driving environments, prediction of uncertain actions of other agents, and human-machine interaction ensuring sufficient situation awareness and engagement of the driver. The list of elements discussed here is by no means comprehensive, but highlights important areas of early development focus. As mentioned previously, the journey along the development roadmap will likely provide greater insights and uncover more proposals to be added to the list.

5.1. The Role of the Operator

Humans typically express the need for retaining control (beyond their fundamental legal responsibility), feeling that is safer and more secure than giving an unknown black box full authority over a highly complex task that, with an error, could seriously jeopardize their life or health. Since automation is classically described as better suited for dull, dirty, and dangerous activities, a driver in the autonomous personalized transportation mode will most benefit from Full Driver Assist functions. These functions offload moment-to-moment driving tasks, such as moving the driver from direct control of the throttle, brakes, transmission gear selector, and steering wheel, to predominantly a command mode. The driver then becomes an operator, who is still in charge, but in supervisory mode, like the orchestra conductor who commands all the instrumentalists (stop/start, faster/slower, louder/softer), but does not play the instruments himself. Even though the operator may be less involved in the moment-to-moment, direct control of actuators, the operator will need greater awareness of the situation, system status, and behavioral intent than is currently available to properly supervise the vehicle's actions. Through Full Driver Assist, the driver is provided additional time and can thereby have more confidence in performing a more appropriate role in the overall system, one that is partially tactical but becomes mostly strategic in nature.

Today, the automotive industry is providing driver support systems in private vehicles to help the driver in critical situations. Warnings, followed by preparation of actuators for operation, are used in sequence in an effort to guide the driver towards a collision avoidance response. Even with the best driver support systems, not all human responses will be ideal; some will inevitably be sub-optimal, not taking full advantage of the support system. The industry is therefore beginning to provide limited autonomous emergency actions in an effort to avoid or reduce the likelihood of an imminent collision. Many, if not all systems allow the driver some override capability versus the autonomous actuation, such as steering away to preempt, cancel or counteract an auto-braking function, if that is preferred. In a similar vein, limited autonomous driving support such as ACC has been introduced, with strict limits on control authority (longitudinal control only, limited deceleration levels, warns driver when control limits have been reached). On the other hand, allowing the driver to override the autonomous system would allow the driver to mistakenly override it as well; yet employing this method allows the earlier introduction and benefit of these autonomous systems.

When will we be ready to override human action with machine action? Flight control logic in modern aircraft already limits pilot input authority to a level which the plane's computers determine is within a safe operating regime. However, transportation modes that currently employ higher levels of autonomy vis-à-vis private road vehicles have one thing in common: very limited interaction with other operators. Airplanes are typically spaced a mile apart or more. The tightest train schedules place trains at least a few minutes apart, and the separation experienced on the ocean, without a harbor pilot aboard, can be even larger. This limited interaction significantly reduces the exposure to the unpredictability of the human reaction / interaction. On the other hand, consumers have an intuitive understanding of the complexity of interaction among vehicles sharing a road. This will likely slow their acceptance and adoption of fully autonomous vehicles.

Given that autonomous vehicles will change the very nature of driving, it is conceivable that the licensing of vehicle operators will need to change along with it. Today we have graduated driver's licenses with legal limitations, and as a driver fulfills certain requirements, more capability gets 'turned on'. Driver training today is mostly limited to several hours of on-road instruction, followed by real-world driving practice to build experience.

More specialized training may become the future norm. This training could include education on advanced CADS systems so that drivers will be better equipped to use the more advanced autonomous driving systems, similar to the pilot training required to fly a significantly autonomous commercial airliner. At some point, we may transition the first autonomous systems to only those in the driving public who have undergone specialized training, earning a certification and a special license to operate an autonomous vehicle. Ultimately, as autonomous vehicle technology matures and becomes more common, an even higher level of training and certification may be required to drive a vehicle in the totally manual, autonomous-off mode.

5.2. Communicating with the Operator

The Human Machine Interface is critical to continued operator engagement, and human-centered design will be essential for ensuring the HMI is properly designed for two-way interaction. The system must communicate everything the human operator wants to know in order for them to be comfortable with the autopilot driving the vehicle. Its effectiveness would be enhanced by knowing something about the operator's state as well.

The ultimate HMI for the autonomous vehicle may be the Brain-Machine Interface (BMI), first demonstrated experimentally in 1999.[33] The Full Driver Assist BMI application would benefit from operational feedback, proprioceptive-like cues, but on a vehicle basis. Similar to the notion that an autonomous vehicle will be available in just a few years, recent public demonstrations have combined with the magnitude of BMI's potential resulting in an enthusiasm that outreaches its readiness. Then again, there are many valuable and arguably necessary intermediate steps before that is realized in common practice.

Today's HMI systems focus mainly on general warnings that only give limited directionality and context. Continued research will be required to understand the best warning methods given the technology of the day, typically audible and visual. A recent study showed that haptic indications work well too, acting almost as a subconscious indication to induce mode changing. When warned at a point that a mode change was not expected, i.e. when a warning was given well before a problem arose that would be difficult to respond to, the operator reacted well to the inducement. When warned at the point that a mode change was proper and expected, the operator continued appropriately without distraction.

To enhance the human response, the HMI must evolve from generating warnings to providing a more immersive, situation-aware, experience. Improved situational awareness is important even in today's limited automatic control features such as ACC, where automatic control in benign situations reverts back to human control when the situational requirements exceed the control authority of the system. Emergency handoff, especially without proper context, is ill-suited to human behaviors. Human attention could waver during autonomous control and the operator may not be prepared to take decisive corrective action.

To improve awareness, the HMI could provide continuous feedback. Steering responsiveness or resistance could be altered as the vehicle gets closer to the lane boundary in order to provide feedback on lane position. Sound could be piped in to the operator correlating to the traffic conditions. With more traffic, there could be greater subliminal presence of sound. If a threat is increasing, then perhaps a localized and directional high frequency sound could be provided, getting louder as the threat grows.

Augmented reality displays (e.g. full-windshield Head-Up Display or wearable display) might be employed to provide directionality and improved awareness by highlighting objects of interest or displaying other scenario information. To achieve the even grander levels of autonomy sought by some, insight into HMI designs that allow the driver to take on more tasks, yet still be engaged, would be required. For the dull driving task, the augmented reality display could be supplemented with driver gaze monitoring to provide pertinent information as the driving scenario becomes critical, when the operator needs to be focused back onto the road. Warnings would still have their role as the last resort, but given an immersive situational awareness the driver would be more involved, informed and active in his role, so when it is time to hand over from autonomous to human control it's not a surprise, the context is understood and it will be a mutual decision. The autonomous system could request confirmation of readiness or willingness for handover of control. This request could be orchestrated so as to preserve a fall-back option of transitioning the vehicle to a non-moving and safely-positioned state suitable for an indefinite period of time (e.g. park it at the side of the road) if the driver doesn't respond or chooses not to accept handover from the autonomous control.

Another goal for a more advanced HMI would be to ensure greater awareness of *evolving* threats such that multiple simultaneous threats can be understood and prioritized, minimizing the need to respond to more than one at the same time, by dealing with the most critical earlier than necessary. In the meantime, other threats could mature or diminish, but all would be strung out sequentially and dealt with before any become critical for response, much the way an air traffic controller would handle it.

As mentioned previously, the autopilot may also need to determine whether or not to rely upon the interruption and guidance of the on-board human. For example, if the driver is in a sub-optimal awareness state (e.g. intoxicated), the computer may need to pursue a completely different task, such as preventing the operator from starting the car. The machine should also protect for the situation where the driver is in perfect operating condition, but misjudges the situation, such as when estimating the closing velocity of a vehicle (something that humans have difficulty doing), not seeing the 2nd car in the line of traffic, missing the car approaching from the right when looking to the left, etc. As the capability is developed, the HMI should include both direct and indirect driver monitoring and interpretation of operator state to ensure properly coordinated driver assist.

The transition from 'driver' to 'operator' will likely take decades, but it has already begun as previously discussed. Tomorrow's HMI designs should help guide and nurture this transition, but large step changes in HMI design may slow consumer acceptance. Therefore designs should evolve smoothly and gradually. Before the autonomous personalized transportation system is realized, the semi-autonomous systems (e.g. CADS) must gradually raise driver familiarity and comfort level for the warning, control, support and interventions of partial automation.

5.3. Deriving Situational Awareness

Real-time, up-to-date information is another critical element of the system. This includes information about the dynamic states and intended action of other vehicles; road hazards, environmental information (including weather, road conditions, natural disasters, etc), or road infrastructure information (e.g. traffic lights are not functioning ahead). The types and amount of information available to road vehicles today lack the reliability and comprehensiveness required to meet the demands of an autonomous personalized transportation system. It is improbable to think that these systems alone could predict other non-autonomous vehicle intentions or their likely future state, and little help is currently available from infrastructure-based information flows.

The radars, cameras, GPS/INS, and map data implemented in today's vehicles are key building blocks for the future; and many more advances are in the foreseeable future. Monocular vision systems may lead to stereo. Lidars may reappear in earnest with scanning multi-beam designs. Flash lidars or 3D cameras may mature enough to enable low cost long-range sensing providing dense range and intensity maps with integrated night vision capability. The numbers and coverage of these sensors will expand to encompass 360 degrees around the vehicle, with longer range and improved positioning and classification.

Additionally, sensors are needed to determine vehicle position relative to proper path. Current localization methods, however, are not precise at all times. For example, GPS positioning accuracy may fall below necessary levels due to atmospheric inconsistencies, drop out zones (due to a tunnel, tree canopy, etc.) or multi-path (urban canyons) failure modes. Alternatively, localization through a comparison of geographic and infrastructure artifacts detected by an on-board sensor to self-generated or publicly available 3D maps may also become important. This technology was demonstrated during the DARPA Grand Challenge 2 and improved in the Urban Challenge Event; subsequent study suggests capability with a single beam scanning lidar within centimeter levels of accuracy. Moreover, 3D maps are on their way, with a number of companies recently discussing their development publicly.

Vehicle-to-vehicle (V2V) and vehicle-to-infrastructure (V2I) network communications can be considered a sensing element that will significantly improve the accuracy and timeliness of information when fused with other on-board environmental sensing. V2V and/or V2I communication (V2X) will enable visibility of other vehicles when direct line of sight is blocked. It will also enable new information to be passed to vehicles, including traffic, weather, and road conditions, and information about the states of other vehicles. Infrastructure information may include environmental sensing of the road network through sensors on the roads, such as placing lidar localization targets in areas with GPS blackouts, or through compilation of the on-board sensing data available from other vehicles connected to a V2V network. If the detection or prediction of low mu conditions prior to encountering them is not yet possible, communicating the experience of a preceding vehicle to others approaching the hazardous area by V2X is a good alternative. The information update and flow would need to be seamless, not only from vehicle-to-vehicle, but also to/from the government, industry, and private sources. New invention and coordination is necessary to make sure the data is the most recent and relevant to autonomous personalized transportation vehicles.

Ultimately, sensing will need to evolve to 'general case' detection, tracking, and classification. Sensors today interpret the world by looking for patterns that match known objects, some of which use a training set and classifiers. Automotive radars are designed to look for a vehicle, which is why they initially worked only on faster moving objects in the driving scene. On the other hand, when humans see the world, they

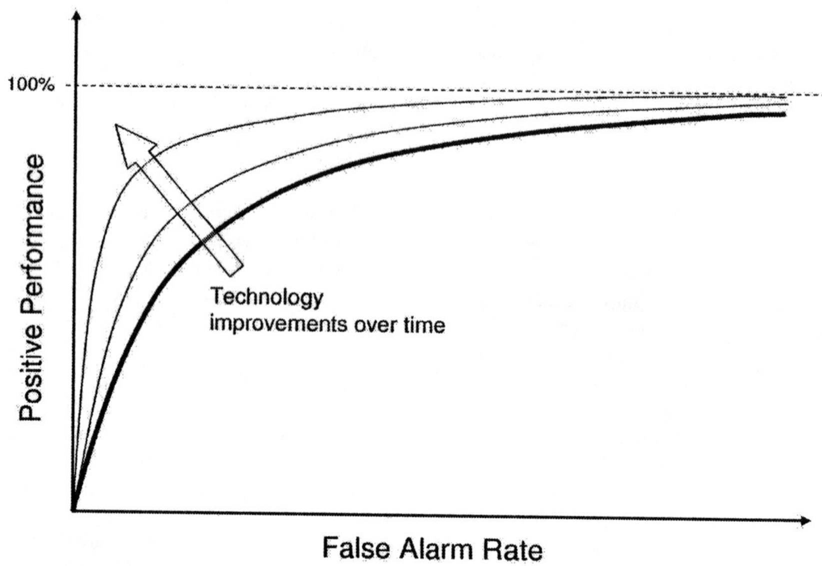

Figure 1.

also look for other cues that help determine whether or not the object ahead is of interest, or if the road is safe to traverse. Beyond just a measurement, there is a level of interpretation and judgment that must be implemented with the sensing system. This would allow estimation of lane and road boundaries when they are not really visible, due to faded, snow covered, glare-obscured conditions or judgment that an object in front, be it a vehicle, bicycle, pedestrian, tree, or moose, may be of interest; or even the gut feeling humans get that the scenario ahead may become a threat and the system should be wary. Knowing that sensors can physically measure much more accurately than humans, we should strive not only to replicate the human sensory perception capabilities, but also to exceed them. An important aspect of this is the use of multiple modalities of sensing in order to address the important problems of sensor reliability and validation of the sensor readings. The common sense verification mechanism that naturally accompanies human perception should be replicated in autonomous vehicles as algorithmic preprocessing validation of the measured data and capability for inferring and predicting new events through associative and case-base reasoning.

5.4. Limits of Situational Awareness

Sensors for situational assessment or awareness (SA) are statistical in nature, merely returning a digital representation of the external environment that must be interpreted for accuracy. Not only do the accuracies of the target characteristics have to be interpreted (e.g. relative range, range rate, and azimuth as well as classification, etc.), but whether the detection itself is valid also needs verification. Both radar and vision systems provide ample targets for interpretation. So it becomes a matter of trading off the true vs. false detection rate (i.e. positive performance vs. false alarms for a collision warning system) for a given modality and specific hardware capability, and then tuning along the curve for an appropriate level of reliability and robustness as shown in Figure 1. As SA technology improves, the tradeoff relationship improves, thereby shifting the curve. This is not much different from when the human acts as a SA system, with cognitive systems that include inductive reasoning, which by their nature, occasionally reach erroneous conclusions even when the basis for it is true.

Humans will never attain perfection, yet we allow them to perform challenging activities, tacitly accepting the consequences. How much better does a machine have to be than the human it would replace, before society allows that replacement to happen? Without knowing the answer, we can still utilize the machine as a situational awareness tool, not feeding an autonomous decision and control system, but in a limited capacity as a driver's aid. Machines are less susceptible to distraction so can provide a benefit given their greater diligence alone. Perhaps it is not a matter of how good an SA or decision-making machine is, but more a matter of how well it learns. Maybe it will be sufficient to allow replacement when it performs *and* learns at least as well as a human, i.e. without making the same mistake twice. Perhaps to break through into a truly autonomous decision making machine, it must be required to, even designed to, learn from and not repeat the mistake of other machines that previously made such an error? The industry has much development ahead before making that determination, but future SA systems should be conceived with consideration of these limitations in mind.

Perhaps the single greatest challenge to effective situational awareness is the speed at which the vehicle must travel to be considered a valued mode of transportation. Initial robotic successes were characterized by the very slow, seemingly deliberate, pace at which the sensing platform traversed the environment. With increasing velocity comes a need for increased sensing range, speed of situational interpretation, hazard detection, classification, and path planning, as well as reliable dynamic control.

5.5. The Vehicle and Artificial Intelligence

The artificial intelligence (AI) that commands the autonomous control system must also evolve, but the evolutionary path is still unclear. Should it be nondeterministic, implementing stochastic type algorithms of learning, optimization, decision making, planning, and goal formation under different situations and states that are not generally known in advance? We don't really know how useful that will be in the long run, but that may be a function of how strong the match must be between the pre-programmed and actual event. Does it need to be more human-like to be self-sufficient, being intuitive, adaptable, and strategic in its functionality? On the other hand, it is important to remember human fallibility; we're not even sure yet how much involvement the operator should have in the system.

We can say that whatever the AI, it needs to handle some level of unexpected environmental perturbations, because chaos exists even in a tightly controlled system. The AI needs to handle any intentional system compromise, for example, dealing with external hacker attacks and false signals. It needs to handle unknown objects in the external environment, like a new type of vehicle on the road that doesn't communicate. It needs to handle unexpected internal failures such as electronics and software faults. The AI really needs to make use of information whenever and wherever it's available, making judgment as to which information to use and when.

Moreover, the AI needs to be able to make decisions spanning both physical safety and societal norms, accounting for the social, political, and cultural complexities inherent in human decision making. Even in a task as simple as a lane change, the decision making logic is complex. When is it safe to make a lane change? When is it appropriate to make a lane change? When is it socially acceptable for an autonomobile to make a lane change? Is it ever acceptable for one autonomobile to cut in front of another, say in an emergency? And in mixed mode operation, one driver may feel comfortable handing control over to his autonomobile, but are other drivers in the adjacent lane ready? All this presumes learning specific driver's actions and preferences in the operation of the vehicle. The models are later used by the intelligent control system to invert the mapped relationships and advise the driver for the most appropriate actions under specific circumstances. All these questions impose requirements on the AI system that are well beyond the capability boundaries of the existing decision making systems and suggest a wide range of challenging research problems.

5.6. The Road Infrastructure

Infrastructure may also require modification to support future autonomous operational modes. As we transition towards full autonomy, we must accept that mixed mode operation may be the norm for a long time, with both human and computer pilots interacting on the road. Some thought needs to be given to this transition - given the uncertainty of human reaction and the interactions that result in random events, we may look to minimize this uncertainty by some day providing special autonomous-only traffic lanes, much like the High Occupancy Vehicle carpool lanes demarked <HOV> today. These lanes could have very limited access, with known access locations, allowing only autonomous pilot-enabled vehicles to enter.

When enough vehicles on the road have autopilot capabilities, we may progress to having some roads, such as limited access highways, be autonomous only; while human drivers could still operate on secondary roads. Eventually, we may transition to virtually all roadways being autonomous only, with only a few exceptions, such as scenic Route 66, preserved for nostalgia's sake.

5.7. The Regulatory Environment and Beyond

While government and regulatory environments will need to adapt to enable the autonomous future, and will likely play a key role in their success, non-regulatory ratings can drive OEM strategies with the same rigor. These latter ratings include government ratings such as NHTSA's New Car Assessment Program (NCAP), as well as third party ratings such as the Insurance Institute for Highway Safety's Top Safety Pick. Many vehicle manufacturers emphasize their performance on these ratings as a communication strategy for vehicle safety; hence these ratings have considerable clout and could even be considered defacto regulations.

Collision avoidance technologies are the fundamental building blocks for autonomous vehicle operation and have been subject to 3rd party influence since NHTSA's NCAP action in 2002 (which applied the fish-hook performance test criteria to ESC systems) which was followed by EuroNCAP braking requirements in 2006. These actions have reverberated around the globe, with Korean, Japan, and China NCAPs all enacting dynamic rollover requirements.

Based on recent history, some NCAPs evolve into regulations. In the preceding example, the US began mandatory phase-in of requirements for ESC by the 2009 model year, a 14 year lag from introduction to regulation. In contrast, regulatory phase-in of passive restraints, a combination of automatic seatbelts and airbags, began in 1986, while a full phase in of airbags began in 1996. A shorter delay is not necessarily preferred even though it can create an earlier 'pull'. A longer delay provides more time to evaluate different technologies and let them mature.

This path is not universal with respect to steps or timing either. In 2010, the US launched a new NCAP Assessment for collision avoidance, with the addition of a FCW and LDW protocol and test methodology. Just prior to that, Japan elected to proceed directly down a regulatory path for collision avoidance, kicking off "if fitted" requirements for CMbB systems, as well as convenience based technology like ACC and Reverse Parking Aid systems. EuroNCAP also just announced the "Advanced Award" (formerly referred to as Beyond NCAP) to supplement the overall safety star rating of the vehicle if the vehicle has Blind Spot, Driver Distraction, or Lane Departure Warning capabilities or Advanced Emergency Braking Systems (AEBS). This can result in near-instantaneous rating assessment of the newest technologies.

These are likely just the first stages of many more requirements to come. Industry is closely watching the US and the EU for regulatory movement in collision avoidance beyond stability control. The US Crash Avoidance Metrics Partnership is a collaboration between several OEMs and NHTSA, researching crash imminent braking system test methods and requirements, among other things, which may result in new NCAP or regulatory requirements. The European Union has already begun to shape commercial vehicle regulations for AEBS and LDW systems, with the United Nations Economic Commission for Europe planning to develop technology requirements in the near future.

Many in the automotive industry are looking for harmonization of these new requirements, with the hope that ISO standards, which exist in either a released or draft form for many of these new features, become the foundation. If harmonization attempts are unsuccessful, the OEM base will face a substantial challenge as it drives toward global technology platforms. Regionally unique requirements could result in key enabling technologies that are unique at a fundamental level. Considering the preceding SA tradeoff discussion (Section 5.4, Limits of Situational Awareness), this could result in one market having a stringent false positive reliability requirement, while another elects to have a high degree of positive function capability, and a third market implements a more simplistic feature presence-based rating or regulation.

Make no mistake, governmental action can stimulate and encourage development of technologies, especially in infrastructure intensive areas, but it should also be careful to not regulate in ways that are restrictive to innovations with societal benefit. All things considered, however, CADS and autonomous vehicle research and development could greatly benefit from the inclusion of governmental agency and legislative partnerships.

6.0. NEW COLLABORATIVE RELATIONSHIPS

Several key factors affecting the pace and extent of innovation are the generation of new concepts, available investment levels, and available time to mature them to a meaningful implementable level.

The solution to complex problems such as Full Driver Assist can only come from the synthesis of many diverse inputs, from diverse sources, and through cooperative relationships. The large investment that will be required presents its own challenge, and that burden is well suited to collaboration as well. Achieving new goals typically requires new skills, developed on the job or gained through additional education, yet both require significant time. Alternatively, skills can be immediately brought into the team by partnering outside your own enterprise.

The traditional supply base is focused primarily on solving today's problems; that is where the majority of demand is, where their expertise is, and where they can be profitable. Yet suppliers also earmark a portion of their budget for R&D to solve future problems. How to spend that investment is a challenging question, with some suppliers extending today's knowledge and others branching out in new directions. Maintaining a regular dialogue with suppliers on trends and new directions ensures alignment and efficiency, but gaps can arise when there is a discontinuity, such as that presented by Full Driver Assist. Sometimes disruptive (i.e. beyond evolutionary) technologies, whether they're from traditional or non-traditional sources, are required.

Disruptive technologies may come from traditional suppliers, but also from other industries, percolating from advanced engineering, fundamental university research, or wherever inspiration may arise, even nature. This opens the door to new entrants in the technology supply base and all should be considered. Looking in non-traditional areas can be like early gold prospecting; you eventually find what you were looking for, but you would probably dig a number of empty holes first.

The following is a partial outline of collaborative relationships that have been or are being explored, but they are presented in a generic and partially fictionalized way. For the purposes of this paper, it is less important to discuss a specific set of corporate relationships, and more relevant to illustrate the breadth and variety of partnerships and technologies, both traditional and non-traditional.

6.1. Traditional partnerships

6.1.1. Tier 1 and 2 suppliers

Long standing chassis and body electronics suppliers are essential contributors to the rapid development and proliferation of new collision avoidance and driver support system technologies. They have proven their capability through the years, but now their out-of-the box creativity is being tested. An opportunistically timed new feature or functional capability breakthrough has the potential to extend their market share overnight in a highly competitive and otherwise mature market.

6.1.2. Pre-Competitive OEM Partnerships

Most notable in this category is the Crash Avoidance Metrics Partnership (CAMP), a research consortium of automobile manufacturers and suppliers engaged with the United States Department of Transportation for the advancement of promising new active safety technologies. This has been a highly effective and productive relationship, having generated numerous concepts, requirements, specifications, and field operational test results on track for eventual implementation.

CAMP's role in the development of V2V and V2I safety communications could serve as a model for Full Driver Assist. Since 2002, CAMP has organized multiple OEMs to work cooperatively on this technology with NHTSA and other parts of the US DOT. The work has ranged from basic testing and analyses to building applications to developing necessary standards and then working together to get these standards adopted. The OEMs currently working together at CAMP (Ford, GM, Honda, Hyundai/Kia, Mercedes, Nissan, Toyota and VW/Audi) are completing the standards necessary for a NHTSA deployment decision in 2013. To support this NHTSA decision, the OEMs working together at CAMP are also building vehicles with this technology for Driver Acceptance Clinics and for model deployment.

To support full commercial deployment of V2V and V2I safety communications, OEMs and the government needed to come together to define the enabling pre-competitive elements, such as infrastructure requirements, as well as message protocols, content, and security, etc. OEMs will need to be able to trust the wireless messages that their vehicle receives from vehicles manufactured by their competitors to provide warnings to the drivers of their vehicles. The level of cooperation and trust for Full Driver Assist applications will need to be examined and, if appropriate, mechanisms such as CAMP should be utilized.

6.1.3. Academia

Also common are relationships with colleges and universities ranging from a one-time grant to formal multi-year alliances. These can in turn leverage research funding from governmental science and military sources, industrial military sources, health care providers, etc. as well as collaborative relationships with other universities.

One quickly finds that university faculty, students, research staff, and affiliated technical institutes working in areas directly relevant to Full Driver Assist form a rather small community, yet draw upon knowledge, skills, and experience from non-automotive ground (construction, agricultural, industrial) and marine vehicles, general/commercial/military aviation, planetary exploration applications, medicine, and brain & cognitive science.

6.2. Non-Traditional Partnerships

Non-traditional partnerships are especially important in tough economic times. You can readily find a partner on a pay-to-play basis, but you easily exceed tight budgets with aggressive long term research when there is a priority on near term results. Non-traditional partnerships often arise when both partners have budget challenges and are motivated to find an equal equity partner, one that brings intellectual capital to move new concepts forward. These can be very strong relationships when they are born from mutual dependence, toward a shared ultimate goal/vision and well aligned with individual goals. The title for each of the following examples serves to capture the essence of these unique relationships.

6.2.1. The Mental Athlete

Formal contests, or any competitive context, can provide motivation and a means for a technical staff to perform at very high levels of creativity on a very short time scale. These contests are common in academic circles and range from toothpick bridges, baking soda cars, and science fairs for the younger set, to high performance and fuel-efficient ground vehicles, concrete canoes, and energy and space efficient homes for those more learned.

This approach to innovation is especially powerful when the team constituents are multi-disciplinary and blended from academics, OEM, suppliers, etc. This has likely driven the recent expansion to include competitions aimed at motivating professional participants as well. These competitions investigate topics ranging from human powered flight, to commercial space flight and space exploration, to ultra-high fuel efficiency, education, health care, and beyond.

Those well suited for this high energy, high stress, instant feedback, creative environment can find themselves supporting professional competition or time sensitive high-stakes consulting teams (e.g. Formula 1 racing, or oil rig fire control, mine collapse rescue, etc.). The downside is that this high level of energy is difficult to sustain for indefinite time periods, and can result in burn-out if continued for too long.

In the Full Driver Assist context, the most notable examples have been contests sponsored by the Defense Advanced Research Projects Administration (DARPA), namely their two Grand Challenges and their Urban Challenge for autonomous vehicle operation. These have drawn hundreds of teams from around the world and brought the notion of 'driverless cars' into mainstream media with widely publicized demonstration events, all while technical advancements (primarily software) are finding their way into further research activities behind the scenes.

6.2.2. The Start-up

Every once in a while a group of engineers has an idea that is ahead of its time, at least within their current context, which warrants a parting of the ways. This has happened several times in the robotics community, and in one case, the engineers decided to spin themselves off from their military contractor parent and start their own company, rather than bookshelf their ideas. Specializing in situation awareness, path planning, threat assessment, vision/image processing, proprioception, search/processing prioritization, and real-time computing, these individuals are highly regarded in the robotics community, regardless of their venue, and they have made good on their vision.

An OEM seeking to push the envelope can learn from such an organization, working together to explore different theories and rapidly prototype complex sensing and control systems with great utility. Their story ends with their former parent organization re-recognizing the value of their abilities, accomplishments, and vision, and ultimately reacquiring them.

Another form of the startup, graduating university students, is also common and possibly more predictable. Typically graduate and undergraduate work is extended into a focused product or services business model by those funding their research. This presents a ground floor opportunity and can be especially powerful if they're also building upon a Mental Athlete collaboration model - first hand knowledge and proven under fire.

6.2.3. The Hobbyist

How often does it happen that someone turns their hobby into a new business and becomes a new entrant in a highly competitive field? It only has to happen once, in the right technology, and you have the makings of a potent collaboration - if you are in on the ground floor.

In one case, a hobbyist applied curiosity, a little inspiration, and a lot of perspiration to develop a new sensing device. This device wasn't entirely novel, but it was uniquely capable nonetheless. It solved a much larger portion of the general case SA problem than had previously been accomplished, addressing road departure and safe path detection, planning, advice, and control.

This sensor is currently being used as an instrument grade research tool and is being produced at low volume for architectural applications, among other things. It has put incumbent sensor suppliers on notice, illustrating that there is a disruptive technology opportunity. Perhaps with additional packaging, manufacturing, and robustness development, this technology will become suitable for automotive applications.

6.2.4. The Gamer

They may 'only' write software for video games, but a serious skill set may be overlooked without a little more investigation. The gamers are really solving an image-processing problem, in their own unique way in some cases, and it is that diversity of knowledge, concept, and approach that can be leveraged. If you find a connection and can draw out their best efforts focused on your problem, the progress could be quite amazing.

6.2.5. The Coach

If you want to teach someone (or an intelligent vehicle) to drive, you might start with someone who is a professional driver, or even better, a professional driving instructor or coach. You, or the intelligent vehicle, need to get that seat-of-the-pants/'been there done that' experience, but without repeating their entire driving history. You need someone to distill and convey it to you efficiently and effectively. Furthermore, advanced driving skills are perishable for humans, so coaching isn't necessarily a one time event.

You (the intelligent vehicle) need to learn the vehicle's nominal character, its limitations, and how it behaves beyond its limits. If this could be done online or in a virtual environment, it could be done in a repeatable way, without the peril of hazardous situations, and in a concentrated fashion. This leaves out the nominal driving mileage and focuses the time on key events and experiences. This might ultimately enable novice drivers to start out with the wisdom of a mature driver, and an intelligent vehicle might embody the natural understanding, presence, and anticipation of a professional.

6.2.6. The Improviser

You need a test method to characterize a collision scenario in a repeatable way, without harm to the test drivers or test vehicles, and you need to ultimately validate such a system. Enter the Improviser. You tell him/her your story and before you know it, something has been discovered in the barn, the hangar, or the tool crib that with a bit of blacksmithing, a few extra wires, and a handful of plastic wrap, perfectly fills the bill. You don't teach someone to do this; this type of person just happens.

6.2.7. The Biologist

The application of chaos and complexity theories in the field of biology is not new, but their application to the human driving condition is. There are inhabitants of planet earth that are wired differently than humans: insects can perform collision avoidance on a time scale, within physical proximities, and with innumerable distractions and clutter, that a professional athlete or intelligent vehicle would be envious of. To understand how to mimic and embed the instinctive as well as cognitive processes observed in nature in future intelligent vehicles, you would do well to diversify your automotive team with this atypical skill set.

7.0. SUMMARY/CONCLUSIONS

It is fanciful to consider practical Full Driver Assist capability achievable in the near or even midterm. Amazing capabilities have been achieved and demonstrated in the carefully controlled environment of the test track, even in the glare of the TV lights. But are we ready to turn this loose on the mainstream consumer? Ultimately the argument of when, or even if, we will ever be ready is moot, as the benefits from the journey itself is worth it regardless the answer.

Having provided a summary of the current challenges and a roadmap for future work, it is fitting to revert to history for some perspective. It has been said that we put mankind on the moon in one giant leap. President Kennedy set forward a visionary challenge and in less than a decade we were there. Why? "We set sail on this new sea because there is new knowledge to be gained ... and used for the progress of all people."

Necessity drove a search for solutions in all conceivable places, the usual and the unusual, but the first moon walk was achieved through a set of logical extensions of what mankind knew. Many challenges remain - more than forty-five years later we still don't have regular commercial service to the moon, earth orbit, or even the upper atmosphere. While our undertaking may not be as grand as putting a man on the moon, perhaps our task is more difficult - there is no road rage in space.

REFERENCES

1. Federal Highway Administration, "The Dream of an Automated Highway," ttp://www.fhwa.dot.gov/publications/publicroads/07july/07.cfm, July/Aug 2007 Vol. 71 No. 1

2. Annual Report, Department of Electrical and Computer Engineering, Ohio State University, 2003.

3. The Great Robot Race, McGray, Douglas, Wired, December 2003.

4. A Perspective on Emerging Automotive Safety Applications, Derived from Lessons Learned through Participation in the DARPA Grand Challenges, McBride, J., Ivan, J., Rhode, D., Rupp, J., Rupp, M., October 2008. Journal of Field Robotics, Volume 25, Issue 10, pp. 808-840. ISSN 1556-4959.

5. NHTSA, "Traffic Safety Facts; 2008 Traffic Safety Annual Assessment----Highlights," http://www-nrd.nhtsa.dot.gov/pubs/811172.pdf, June 2009

6. Reichspatent Nr. 169154, Verfahren zur Bestimmung der Entfernung von metallischen Gegenständen (Schiffen o. dgl.), deren Gegenwart durch das Verfahren nach Patent 16556 festgestellt wird, November 11, 1904. Translation: *Methods for determining the distance from metal objects (ships, or the like), whose presence from the procedure under 16556 patent is found.*

7. Delphi, "Manufacturer Products: Safety Electronics, Active Safety, Delphi Adaptive Cruise Control," http://delphi.com/manufacturers/auto/safety/active/adaptive-cruise-control/, 2009.

8. Smyth, Louise, "Adapt to Survive," Vision Zero International: 27-30, June 2010

9. Stupp, E. H., Cath, P. G., and Szilagyi, Z., "All Solid State Radiation Imagers," U.S. Patent 3 540 011, Nov. 10, 1970.

10. Volvo Car Corporation, "Pedestrian Detection with full auto brake - unique technology in the all-new Volvo S60," http://www.volvocars.com/en-CA/top/about/news-events/Pages/press-releases.aspx?itemid=17, March 2010

11. Frost & Sullivan, "Active and Passive Safety Systems in the US: Customer Desirability and Willingness to Pay (Passenger Vehicles)," Sept. 2005.

12. Frost & Sullivan, Analysis of North American Market for Advanced Driver Assistance Systems, December, 2009.

13. Frost & Sullivan, Strategic Analysis of Japanese Passenger Vehicle Safety Systems, June, 2010.

14. World Health Organization, "The Global Burden of Disease, 2004 update", WHO Press, Geneva, Switzerland, ISBN 978 92 4 156371 0, 2004.

15. Research and Innovative Technology Administration, John A. Volpe National Transportation Systems Center, "Focus," http://www.volpe.dot.gov/infosrc/highlts/02/mayjune/d_focus.html, May 2002.

16. Dekker, Sidney, The Bad Apple Theory pp. 1-14, The Field Guide to Human Error Investigations, 2002. TJ International, Padstow, Cornwall, Great Britain. ISBN 0-7546-4825-7.

17. Dismukes, R.K., Berman, B. A., Loukopoulos, L. D. "The Limits of Expertise", Ashgate, United Kingdom, ISBN: 978-0-7546-4964-9, 2007.

18. Rothe, P. J., James, L., Nash, J., Freund, P., Martin, G., McGregor, D., Frascara, J., Redelmeir, D., "Driving Lessons, Exploring Systems that Make Traffic Safer" University of Alberta Press, Canada, ISBN-10:0888643705, 2002.

19. Central Japan Railway Company, "About the Shinkansen", http://english.jr-central.co.jp/about/index.htrnl, retrieved: Aug 2010.

20. Car and Driver, "How to Drive a Ford Model T," http://www.caranddriver.com/features/09q3/how_to_drive_a_ford_model_t-feature, July 2009

21. Langer, G., "Poll: Traffic in the United States; A Look Under the Hood of a Nation on Wheels", ABC News, Feb 2005.

22. NHTS, "Summary of Travel Trends: 2001 National Household Travel Survey," http://nhts.ornl.gov/2001/pub/STT.pdf, Dec. 2004.

23. Song, C., Qu, Z., Blumm, N., Barabási, A., "Limits of Predictability in Human Mobility", Science 327(5968): 1018-1021, Feb. 2010, DOI: 10.1126/science.1177170.

24. Rokeach, M., "The Nature of Human Values", Free Press, New York, ISBN 0029267501, 1973.

25. Shrank, D. and Lomax, T., "2009 Urban Mobility Report", Texas Transportation Institute: The Texas A&M University System, Jun 2009.

26. Massachusetts Institute of Technology, Center for Transportation & Logistics, "Driver Wellness, Safety & the Development of an AwareCar," http://agelab.mit.edu/system/files/file/Driver_Wellness.pdf, Dec. 2009.

27. Ford Motor Company, "Ford and MIT Team Up To Improve Safety by Reducing Driver Stress", http://media.ford.com/article_display.cfm?article_id=31682, Dec. 30 2009.

28. Joint Cognitive Systems: Patterns in Cognitive Systems Engineering, Woods, David D., Hollnagel, Erik. Published by CRC Press, Boca Raton, Florida, 2006. ISBN 978-0-8493-3933-2.

29. Cacciabue, P., Modeling Driver Behavior in Autonomous Environments, Springer-Verlag, London, 2007.

30. Lu, J., Filev, D., "Multi-loop Interactive Control Motivated by Driver-in-the-loop Vehicle Dynamics Controls: The Framework," Proc. of the IEEE Conference on Decision and Control, December 16-18, 2009, Shanghai, China.

31. Pentland, A., Liu, A. (1999) "Modeling and Prediction of Human Behavior", Neural Computation, Vol. 11, pp. 229-242.

32. Ford Motor Company, "Ford's new smart intersection 'talks' to cars to help reduce collision, fuel-wasting congestion", http://www.ford.com/about-ford/news-announcements/press-releases/press-releases-detail/pr-ford26rsquos-new-smart-28610, July 10, 2008.

33. Chapin, J.K. et al. (1999) Real-time control of a robot arm using simultaneously recorded neurons in the motor cortex. Nature Neuroscience, 2, 664-670.

CONTACT INFORMATION

Jeffrey D. Rupp
Ford Motor Company
Product Development Center - Safety Engineering
Room GB-A83, MD 129
20901 Oakwood Blvd
Dearborn, MI 48124-4077
jrupp@ford.com

ACKNOWLEDGMENTS

The authors would like to thank those who have provided invaluable input and critique for, as well as the investigation and advancement of the technologies which are the subject of this paper: Tom Pilutti, James McBride, Dimitar Filev, Louis Tijerina, Mike Shulman, Alex Miller, Dev Kochhar, Andy Sarkisian, Venkatesh Prasad, Stephen Rouhana, Doug Rhode and Roger Trombley. The authors also gratefully acknowledge those who have facilitated the process of creating this paper in innumerable ways: Nancy Burke, Haleh Ghalambor, Stephen Kozak, Randal Visintainer, and Loralee Rupp.

Major Design and Test Collaborations

SAE The Engineering Society For Advancing Mobility Land Sea Air and Space INTERNATIONAL

2011-36-0060

Sartre – Safe Road Trains for the Environment Reducing Fuel Consumption through lower Aerodynamic Drag Coefficient

Arturo Dávila and Mario Nombela
Applus+ IDIADA

ABSTRACT

The SARTRE Project is an FP7 European collaborative project funded by the EC with the participation of 7 entities from 4 countries. The SARTRE project focuses on the design, set up, evaluation and validation of road train systems that do not require modification to the actual road infrastructure, making use of available technologies for in-vehicle applied systems.

The SARTRE project will define a set of acceptable platooning strategies that will allow road trains to operate on public highways without changes to the road and roadside infrastructure therefore enhancing, developing and integrating technologies for a prototype platooning system such that the defined strategies can be assessed under real world scenarios. Also, the project will show how the use of platoons can lead to environmental, safety and congestion improvements.

The project bases the functionality of the platooning system in safety improvements from autonomous control systems to be used in lead vehicles (trucks) and following vehicles (trucks, cars, SUVs). Considering that 87% of road fatalities include contribution from the drivers, automated systems prove to be safer means while striving for more efficient personal transportation.

Smaller time and space gaps between vehicles with the aid of autonomous control will contribute by improving the traffic congestion, to provide the maximum efficiency. This will translate into a delay in the traffic collapse and reduced traffic dynamics on highways. This will also become an important aspect of the project, as a significant fuel efficiency improvement achieved from a reduction in the aerodynamic drag coefficient of the entire platoon is expected. An average 20% decrease has been found for 4 vehicles with a 0.2 vehicle length gap in previous projects.

SARTRE will initially perform a simulation of the most representative platoon configurations to obtain the expected drag coefficient reduction, in order to then calculate the expected fuel consumption savings. Having this data in hand, a real fuel consumption evaluation will be performed while in the validation trials, to make an exact correlation of the calculated data and the real life data. This is a challenge that has been addressed in order to develop new capabilities that can optimize future aerodynamic calculations and reduce the cost of testing for specific make and models and platoon combinations.

The challenge addressed by the simulation relies on the length of the platoon, the requirement to keep the wake effect flowing towards the end of the platoon and the different platoon configurations to be analyzed, having very different vehicle geometries and gaps in the mix. Up to date, the only simulations and tests available were carried out using platoons with identical vehicles and vehicle gaps.

INTRODUCTION

The main overall objective of the SARTRE Project is to encourage an evolutional change in the use of personal transport by developing a safe and environmentally friendly road train. For the safe adoption of platoons, systems that allow the interaction of the platoon with other road traffic on un-modified public highways will be developed. Hence, this project addresses three cornerstones of transportation: environment, safety and congestion improvement.

Figure 1. SARTRE Logo

SARTRE is envisioned as a three year project supported by the European Commission's 7th Framework Programme. The consortium is formed by seven partner institutions from four different European countries. Along the project, each of the partners will provide its expertise on their field of action to achieve the established objectives of the project, obtaining important technological developments combined with a thorough analysis of the human factors and future requirements of platooning.

The scope of the project involves added human factors to the concept of road trains, making it necessary to carry out extensive testing on each of the systems individually and later integrating all of them. These tests will serve as validation trials for the system, paving the way for the introduction of platooning in a medium term future.

THE SARTRE PROJECT

Previously, road trains had been approached with several different scenarios, where most of the solutions required significant modifications to the road infrastructure or even dedicated lanes. Within SARTRE, we strive to accomplish a more accessible platooning system without road modifications and using state of the art technology available in many of today's cars.

Thanks to the increased reliability and the every time reducing cost of electronic equipment and communication systems, it is viable to develop a safe and reliable platoon system. Nevertheless, and due to the nature of the systems, several challenges stall the introduction of platoons amid conventional traffic, including acceptability. This is still an issue that has not been solved and SARTRE plans on providing demonstration of the system to encourage acceptance of the system from all involved parties, be them legislators, manufacturers, users or future platoon fleet operators.

The SARTRE project and its demonstration tasks, through the full understanding of the related issues of platoons on public roads and developing new solutions that help address acceptability issues, aim at showing to the public the environmental, safety and convenience benefits obtained from the use of this system. In this way, the project encourages the modal shift towards vehicle platoons on motorways.

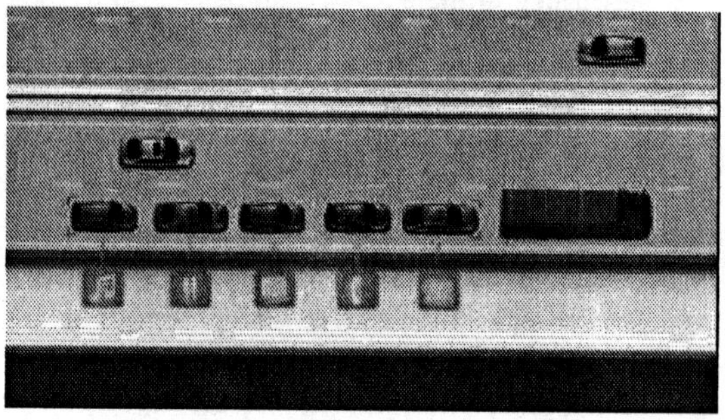

Figure 2. Bird view of a platoon

PLATOONING BENEFITS –

Environmental Benefits

Previous projects have investigated the potential environmental benefits of platoons, that come from a reduction in aerodynamic drag coefficient leading to less resistance and decreased engine output. An example is the PATH Project in the USA, which has been going on for over 20 years, and their researchers have carried out several studies on the aerodynamic performance of platoons, by using wind tunnel tests with scale vehicles, real platoon performance measurement and physical drag coefficient calculation. Their research has proven satisfactory in the reduction of drag coefficient for vehicles in platoons of 2, 3 and 4 vehicles, with several gaps and offsets.

An important literature review on the PATH Project documentation has revealed the following important facts:

- There is simulated and physical evidence of drag reduction for <u>same vehicle</u> platoons.
- The initial calculated data of up to 20-25% savings from calculations, exceed reality, yet the savings have been reported up to 15%.
- All the vehicles involved in the platoon benefit with drag reduction, including the lead vehicle since also pressure variations generated within the platoon help aerodynamically.
- The most benefited vehicles are the ones located in the middle of the platoon, then the trailing or end vehicles, then the lead vehicles.
- The benefit is greater with smaller gaps. Nevertheless, some studies show that from a 2 m to a 10 m gap, there is only 2% difference.
- Having a "clean" lead vehicle (no protrusions or edges, tires, antennas) and a "dirty" trailing vehicle resulted in the best performance. This due to air pressure localized around the vehicles.
- In highways with high exigency on the engines, the air flow coming in for cooling and cabin air is still enough while platooning. Attention would need to be paid for long uphill roads while platooning, although car manufacturers state that vehicles are capable of coping with the exigencies.
- There has not been any simulation or trial using different vehicle shape and sizes, only same vehicle make and model.
- When having an offset, there is also a difference in the drag coefficient, where the most affected car is the immediate vehicle, and the trailing car receives less disruption.

With this in mind, an important work is carried out within SARTRE to establish the optimal gap spacing and running order of the vehicles within a platoon. Initially, a simulation of different platoon configurations is made, where we expect to obtain average drag coefficient modifications from inside a platoon at different gaps.

Generally, a smaller spacing between vehicles gives greater benefit in terms of energy consumption. However, smaller spacing is also more challenging for the control system, so a balance needs to be established. Initially, we were aiming at a spacing of about 1 meter, which would correspond to about 0.2 vehicle lengths, which is supposed to give the most important benefits. Now, we have found evidence from other projects and some own trials, that this is a very short gap which could not give an optimal safety to fuel consumption ratio.

Safety Benefits

A platoon has a different performance on highway traffic, and thus implies different risks for accidents and injuries. There is a clear challenge in having the platoon interact with drivers and surrounding traffic and maintain the capability of safely managing routine and emergency maneuvers. There are obvious benefits of having autonomous vehicles following a professional driver.

A Department for Transport report (see bibliography) states that 95% of road accidents involve human factor contribution. The platoon incorporates a significant level of driving automation whereby for extended periods "drivers" of following vehicles concede their control to the lead vehicle and local autonomous systems. This helps to have much better reaction times and vehicle control during emergency situations, because the systems will actuate to their best performance, whereas a human driver may not be able to react in time and may not perform the adequate maneuvers.

Figure 3. Safety simulation

The system has been designed to have V2V communications that have a very small transmission delay, giving a very low system activation time (for steering, accelerating/braking) which translates into almost instantaneous reactions of the entire platoon. In the case of regular platooning situations, the road train systems are in charge of maintaining the specified performance. In case of an emergency, these systems overcome the natural human reactions and provide an exceptional handling performance, deeming the platoon as a very safe system.

Traffic Flow Benefits

For the expected traffic flow benefits of platooning it is necessary to discriminate between different traffic conditions and the applicable road train scenarios to each of them. There are 4 principal traffic conditions that can be met: free traffic, collapsing traffic, synchronic inhomogeneous traffic and stop&go traffic. Obviously, the best traffic condition is free traffic, from which, according to the increase in volume and variability of gaps between vehicles and speeds, the other three conditions can be met.

SARTRE platooning may not bring improvements for free traffic yet it will indeed aid in the delay of collapsing traffic, maintaining a constant speed and gap between vehicles on a scenario where high speeds and open spaces are a constant. By the time that collapsing traffic scenario is reached, the road train will provide the most benefits, as the point of the collapse is dependent on the required traffic space of each vehicle and the time gap. The smaller the time gap the more the collapse point is shifted towards higher traffic flows. With platoons, this time gap is reduced to a minimum and by doing so the road capacity can be enhanced and traffic congestions avoided or at least delayed.

A latter condition is the synchronic inhomogeneous traffic, characterized by density waves, where vehicles drive in between 30 and 80 km/h in 100 m distance. In this case, a significant improvement can be expected as the autonomous guidance helps reduce the dynamics. In the case of a further increase of the traffic flow a complete breakdown occurs, and stop-go traffic condition appears. The beneficial effect in this condition is achieved when the platoon leaves the traffic jam, as the acceleration is sufficient enough and controlled, maintaining the space between vehicles thus leading to a faster dissolving of the congestion.

SIMULATION – The work being carried out as an initial step for the evaluation of fuel consumption, and hence the expected environmental benefits, is a full simulation using Computational Fluid Dynamics (CFD) software. The simulation uses the different vehicle geometries that the SARTRE Project has established and the tests were selected according to the possible road train configurations we might use during our project.

The results from these simulations will be used to assess which is the better gap between vehicles, offering safety and drag coefficient improvement, and also which are considered to be the best configurations. Having a selection of configurations and optimum gaps will allow us to make a more efficient use of the track time to obtain the physical fuel consumption evaluation and the validation tests.

An added benefit is that we will be able to know exactly what the behavior of the air flow is as it passes over a platoon. With this simulation method, we can arrange our vehicles in many different positions inside the road train and evaluate, for example, whether the benefit is greater for a small car behind a heavy goods vehicles or if having a big SUV behind a compact car will mean that the SUV coefficient will be worsened.

1.-	Truck FH12		Reference Cd		
2.-	SUV XC60		Reference Cd		
3.-	FV1 S60		Reference Cd		
4.-	FV2 V60*		Reference Cd		
5.-	Truck FH12	Following Truck	SUV XC60	FV1 S60	FV2 V60
6.-	Truck FH12	Following Truck	FV1 S60	FV2 V60	SUV XC60
7.-	SUV XC60	FV1 S60			
8.-	Truck FH12	FV1 S60			
9.-	Truck FH12	SUV XC60			
10.-	Truck FH12	Following Truck			
11.-	Truck FH12	Following Truck	SUV XC60	FV1 S60	
12.-	Truck FH12	Following Truck	SUV XC60		
13.-	Truck FH12	Following Truck	FV1 S60	FV2 V60	
14.-	Truck FH12	Following Truck	FV1 S60		
15.-	Truck FH12	SUV XC60	FV1 S60	FV2 V60	
16.-	Truck FH12	SUV XC60	FV1 S60		
17.-	Truck FH12	FV1 S60	FV2 V60		
18.-	Truck FH12	Following Truck	FV1 S60	SUV XC60	FV2 V60
19.-	Truck FH12	Following Truck	FV1 S60	SUV XC60	
20.-	Truck FH12	FV1 S60	SUV XC60	FV2 V60	
21.-	Truck FH12	FV1 S60	SUV XC60		
22.-	Truck FH12	FV1 S60	FV2 V60	SUV XC60	

Figure 4. Possible configurations matrix

There is and enormous amount of information to be generated during the simulation, due to the fact that the air flowing along the platoon will have to be modeled and maintained throughout the length of the platoon, which exceeds easily the 50 meter mark. Another important remark is that in order to have the air flow, we have to simulate wind tunnel that in our case is able to receive the entire platoon. This makes us manage a very large quantity of processing time and require the capability to do so.

For this task, we set out to the CESCA (Centro de Super Computación de Catalunya) to perform the simulation. In here, we find the required processing capacity and the facilities to have the computers working for the next couple of months. In order to accomplish this job, a special script for the automation of the process was written.

The script methodology is as follows:

1) Text input of the 3 variables to a new document.
   a) Order of the vehicles
   b) Spacing between vehicles (gap)
   c) Velocity
2) Pre-process
   a) Automatically, the geometry of the entire platoon is created inside the virtual wind tunnel.
   b) A volume mesh is generated automatically, including the air
3) Process
   a) CFD calculation.- This is the actual aerodynamic modeling.
4) Manual post process.- The results are analyzed.

Figure 5. Render view from the CFD Simulation

The results of the entire process are delivered numerically and visually, through a series of videos showing the flow of air over the platoon.

CONCLUSIONS

By the end of the project, the SARTRE consortium will have created a prototype road train system that will be friendly with the environment, safe for all the users of the road and help in the reduction of traffic jams by the aid of better traffic dynamics.

A very important aspect will be the reduction of fuel consumption. To this means, IDIADA has started a super calculation task where several configurations of platoons will be tested and the aerodynamic coefficient of the road train will be analyzed. The expectations are that a 15% of fuel save can be found in the best cases, whereas there will always be benefits for all the platooning vehicles.

The use of CFD software reduces the amount of time required to know aerodynamic coefficient, giving us the opportunity to simulate the most relevant configurations and deciding if the validation tests will include such cases for actual fuel consumption measurement. Apart from this, it establishes a methodology to further analyze platoon behavior with many different make and model of cars and heavy vehicles, which will be of interest for the future users of road trains.

It is very important to mention that at the moment that this paper is written, the simulation on the CFD is going on automatically al the CESCA in Catalunya. It will take approximately two months to finish the entire calculations of the entire platoon and different cases selected.

REFERENCES

1. Brown, I. (2005) Review of the "Looked but Failed to See" Accident Causation Factor, Road Safety Research Report No. 60, Department for Transport, London

2. INVENT Project website http://www.invent-online.de

3. SARTRE Project website www.sartre-project.net

4. Shladover, S. (2007), "PATH at 20 – History and Major Milestones", IEEE Transactions on Intelligent Transportation Systems, vol. 8, pp. 584-592.

5. Zabat, M. (1994), "Drag Measurements on a Platoon of Vehicles", PATH Project paper.

6. Zabat, M. (1995) "The Aerodynamic Performance of Platoons: Final Report", PATH Project paper.

7. Marcu, B. (1998) "The Aerodynamic Forces on Misaligned Platoons", PATH Project paper.

8. Hong, P. (1998) "Drag Forces Experienced by Two, Full-Scale Vehicles at Close Spacing", PATH Project paper.

9. Mikaelian, M. (2000) "Field Experiments Demonstrate Fuel Savings for Close-Following".- PATH Project paper

CONTACT INFORMATION

Arturo Dávila
adavila@idiada.com

Mario Nombela
mnombela@idiada.com

Applus+ IDIADA
+34 977 166 021

2008-01-2718

Ohio State University Experiences at the DARPA Challenges

Keith A. Redmill, Umit Ozguner, Scott Biddlestone, Alex Hsieh and John Martin
The Ohio State University

ABSTRACT

The Ohio State University has fielded teams at all three of the DARPA Grand Challenge and DARPA Urban Challenge autonomous vehicle competitions, using three very different vehicle platforms. In this paper we present our experiences in these competitions, comparing and contrasting the different requirements, strategies, tasks, and vehicles developed for each challenge. We will discuss vehicle control and actuation, sensors, sensor interpretation, planning, behavior, and control generation. We will also discuss lessons learned from the engineering and implementation process for these three vehicles.

INTRODUCTION

In July 2002, the Defense Advanced Research Projects Agency (DARPA) announced a Grand Challenge [1] for unmanned and autonomous off-road ground vehicle development. Vehicles and development teams were to be entirely self-funded. The vehicles should be able to navigate a course consisting of several hundred miles of off-road terrain in the desert southwest region of the United States, following a path defined by a series of waypoints unknown to the teams until immediately before the race, and negotiating natural and manmade obstacles and terrain features without outside intervention. Once the race began, no contact or communication with the vehicle or its systems was allowed.

DARPA conducted the first Grand Challenge event in March 2004. The course, defined by over 2000 waypoints, would take the vehicle across approximately 150 miles, beginning at Barstow, California and ending at Primm, Nevada. A prize of $1 million (US) would be awarded to the fastest vehicle that completed the course in less than ten hours. Nobody won. In fact, the furthest distance traveled was 7.2 miles.

The second Grand Challenge event was held in October 2005. A 132-mile course, defined by almost 3000 waypoints beginning and ending at Primm, Nevada, traversed rocky trails, dry lakebeds, and mountain passes. This time, 5 teams finished the course (4 within the allowed 10 hours) and the vehicle Stanley, developed by the Stanford Racing Team, took the $2 million (US) prize by completing the course in 6 hours and 53 minutes with an average speed of 19.1 mph.

Unlike the Grand Challenges, the 2007 DARPA Urban Challenge was a race competition to design and implement a completely unmanned, autonomous automobile that can deal with a wide variety of mostly urban conditions and scenarios. Autonomous behavior generation occurred at much higher and more sophisticated levels than in the off-road Grand Challenges, and required sensing and control that could detect and evaluate moving obstacles, understand road situations, plan and replan routes, obey California traffic laws and driving procedures, and deal with stop signs, vehicular traffic, and other obstructions. The DARPA Urban Challenge, although quite complex, did have fairly low speed limits, careful drivers and no traffic lights. Visual lane markings were unreliable, and thus true to life, and the terrain was fairly flat, although some areas were unpaved, generating an unusual amount of dust and creating problems for some sensors.

Successful implementation of a Grand Challenge autonomous vehicle involved a number of daunting requirements:

- Very accurate position and orientation localization, even with GPS blackouts
- Sensing of vehicle environment and state in a complicated, semi-structured environment
- Long term autonomy and vehicle control over an unknown course and terrain
- Long-term robustness of both hardware and software in a bumpy, dusty, hot, and occasionally wet environment
- Safe behavior and performance of the vehicle in the absence of an onboard human driver

- Completion of significant testing and validation efforts

For the 2004 Grand Challenge our team, a partnership with the Oshkosh Truck Corporation, developed Terramax'04 [2,3,4,5], an Oshkosh Truck Corporation Medium Tactical Vehicle Replacement (MTVR) 6-wheeled heavy duty vehicle, as our autonomous off-road vehicle entry. This vehicle is shown in Figure 1. Of 106 applicants, it traveled 6th furthest.

Figure 1: Team Terramax 2004 Challenge Vehicle

In 2005, our "Desert Buckeyes" team outfitted ION: the Intelligent Off-road Navigator [6], based on the much smaller but very robust and capable 6x6 Polaris Ranger utility vehicle shown in Figure 2. Of the 196 applicants, ION traveled 29 miles and 10th furthest.

Figure 2: 2005 Intelligent Offroad Navigator (ION)

In 2007, a 2006 Toyota Highlander Hybrid SUV, shown in Figure 3, was selected for automation as the Autonomous City Transporter (OSU-ACT) for the DARPA Urban Challenge [7]. The interior space available in the vehicle provides significant flexibility for the mounting of processing, power, and control hardware, at the cost of only two additional feet of curb-to-curb turning circle diameter and 4 inches of additional width relative to an ordinary passenger sedan. The exterior has convenient hard points for mounting sensors or attaching mounting brackets. The hybrid vehicle technology provides a number of advantages for automated vehicle purposes. DC power for computers, electronics, and sensors can be derived directly from the vehicle's 280-volt battery pack using DC-DC converters and DC-AC inverters. The task of actuation is also significantly simplified since the vehicle employs electric actuators for its stock systems. The vehicle is inherently throttle-by-wire, as the division of electric motor vs. ICE power is dynamically controlled. Under normal operating conditions the vehicle is essentially brake-by-wire as well: the hydraulic master cylinder is isolated from the actual hydraulic brake system and driver requested braking torque, which is sensed using brake pedal position sensors and pressure sensors in the master cylinder-stroke simulator hydraulic circuit, is measured and then generated automatically through ECU control of regenerative braking and the electrically-operated hydraulic brake system. The transmission is controlled electronically as well, with the exception of the parking gear.

Figure 3: 2007 Autonomous City Transporter (OSU-ACT)

For each of these vehicles, we implemented drive-by-wire capability to control steering, throttle, brakes, and transmission. Various sensors, including GPS, inertial measurement units, multiple digital cameras and image processing systems, lidars, radars, and other sensors were integrated. Additional electrical power generation, equipment enclosures and cooling, and safety systems were also implemented as needed for each vehicle.

SYSTEM ARCHITECTURE

A generic functional architecture for an autonomous vehicle is given in Figure 4. Each block is, of course, tailored to the required functionality and degree of autonomy of the vehicle application. But even for the simplest autonomous vehicle scenario, for example a vehicle capable of driving autonomously in the highly controlled environment of an interstate highway, each of these subsystems would be required. The detailed architecture of our 2007 OSU-ACT vehicle is shown in Figure 5.

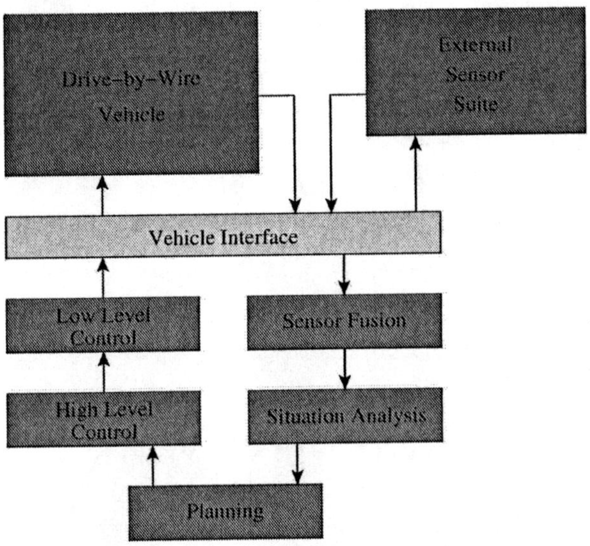

Figure 4: Generalized Autonomous Vehicle Architecture

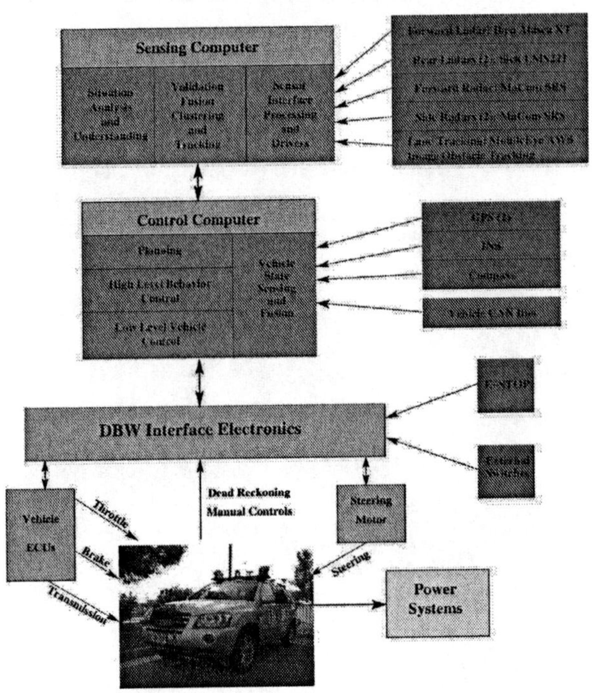

Figure 5: OSU-ACT Architecture

PLANNING

Planning and path generation can take a number of forms depending on the required application. In the Grand Challenges, a starting point, an ending point, and a single route between them was defined by a series of waypoints, latitude and longitude coordinates, along with a corridor width between each waypoint. The vehicle was free to travel anywhere in the corridor. There were no high-level decisions to be made. The desired path could be defined by fitting smooth spline functions through the waypoints, and deviations from this desired path were generated as a reaction to local sensing information.

For the Urban Challenge, however, the behavior was defined as an ordered series of goal locations the vehicle was to attain. The vehicle might start from any location, and the route had to be planned in real time over a map database defining the road network as well as parking lots (zones) and parking spaces. For this task, there were often multiple possible routes and an optimal route had to be identified based on estimates of travel time. The planning software also required the capability to remember blocked or impassible roads so that, if an initial plan failed, a new path could be identified.

HIGH LEVEL CONTROL

High-level control can be thought of as behavior generation, and in our autonomous vehicles is expressed as a hierarchy of finite state machines. At the top level, overall families of behavior appear as independent state machines, which we designate as meta-states. The DARPA Grand Challenges of 2004 and 2005 were both off road races. As such, the only behavior and thus the only meta-state required would be path following with obstacle avoidance from point A to point B [8]. However, since there was no path or lane that could be discerned from a roadway, the only method of navigation is to rely on GPS and INS based vehicle localization [9] and a series of predefined waypoints. Obstacle avoidance techniques were needed, although in the less structured off-road scenario greater freedom of movement and deviations from the defined path were allowed. The Grand Challenge race rules ensured that there were no moving obstacles and different vehicles would not encounter each other in motion. General off-road driving would of course not have this constraint.

Fully autonomous urban driving introduces a significant number of new meta-states, situations where different behavior and different classes of decision need to be made [10]. Figure 6 shows the highest-level meta-state machine that defined the urban driving behavior of OSU-ACT. Each meta-state represents a sub-state machine. These state machines are too complicated to be legibly shown here.

In contrast to the off-road challenges, obstacle avoidance in urban situation presents a much more complicated array of options, including dodging maneuvers within the current lane, lane changes when there are multiple lanes traveling in the same direction, passing an obstacle in an oncoming traffic lane, or even abandoning the current plan, executing a u-turn, and planning a different route to the destination.

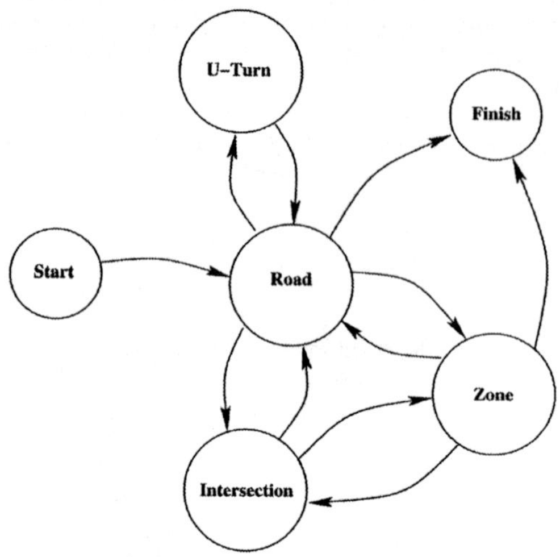

Figure 6: Urban Autonomous Driving Meta-States

SENSORS AND SENSOR FUSION

VEHICLE LOCALIZATION - A key element of autonomous vehicle technology is vehicle localization. All aspects of the system, from sensor processing and fusion to navigation and behavioral decision making to low level lateral and longitudinal control require accurate vehicle position, velocity, and vehicle heading, pitch, and roll information at a fairly high update rate. Providing this information requires the use of multiple sensors, including multiple Global Positioning System (GPS) receivers augmented with Omnistar HP wide-area differential corrections for redundancy, Inertial Measurement Units (IMU), and dead reckoning sensors (wheel speeds, transmission gear and speeds, throttle, brake, and steering wheel position) provided on the vehicle, and a validation system to eliminate sensor errors, especially GPS-related step-change events caused by changes in differential correction status or the visible satellite constellation. To account for sensor errors, noise, and the different update rates of each sensor, an Extended Kalman filter is applied to generate the required state measurements.

EXTERNAL ENVIRONMENT SENSING - There are some distinctions in considering sensing requirements for urban versus off-road applications. We list some noteworthy items below:

- For off-road applications, compensation for vibration and other vertical and rolling motions needs to be done in software or hardware, for example using the IMU and sensor data to specifically generate a "ground plane" that can be referenced while doing sensor validation and fusion. Sensor adjustments are also required to deal with dust, rain, and changing lighting conditions.
- For domains where there are many moving obstacles (i.e. urban applications) one may need to "track" individual obstacles at all times
- Specific operations (parking, dealing with intersections, entering/exiting highways, etc.) may use totally separate sensing and sensor architectures tailored to the task.

The sensors used on ION and OSU-ACT and their effective coverage areas are given in Figures 7 and 8.

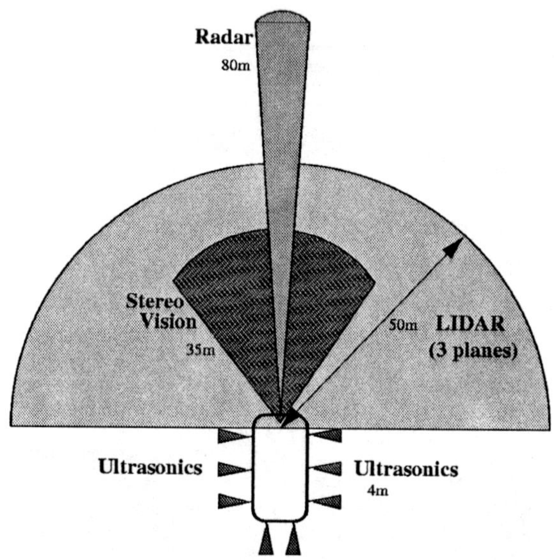

Figure 7: Sensor Suite and Footprint for ION

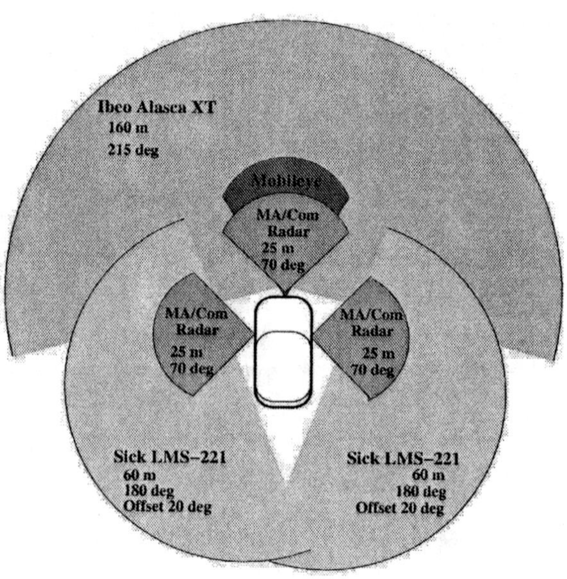

Figure not to scale.

Figure 8: Sensor Suite and Footprint for OSU-ACT

As can be seen in the figures above, the sensor coverage must be tailored to the application. The primary consideration for off-road driving is obstacles in front of and immediately beside the vehicle. Inexpensive short distance ultrasonic sensors were sufficient to allow for a small, slow backup maneuver and to provide side sensing in very tight corridors. In an urban scenario, where sensing was required to support sideways maneuvers into alternate lanes, u-turns, intersection negotiation, and merging into oncoming traffic, sensing for a significant distance in all directions around the vehicle was required.

SENSOR FUSION - In general, we identify two approaches to sensor fusion: a grid or occupancy map approach and a cluster identification and tracking approach.

In the grid map the sensing architecture for sensor fusion is established by developing a cell-based grid map of the vehicle surroundings. All external sensors feed into this map with obstacles sensed and related confidence level. The map is maintained internally in vehicle centered world coordinates and translates but does not rotate with the vehicle. The sensor fusion algorithm implemented on OSU's 2005 DARPA Grand Challenge vehicle used such a grid occupancy approach [11] shown in Figure 9.

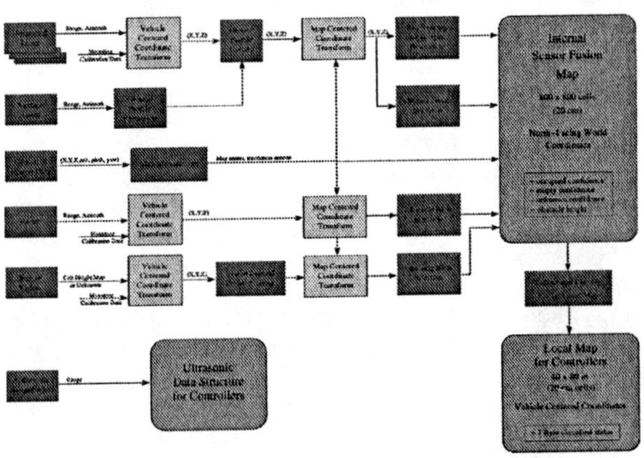

Figure 9: Block Diagram of ION Sensor Fusion

Due to the presence of traffic situations where the environment is highly dynamic OSU-ACT utilized an approach in which the sensor fusion algorithm is responsible for clustering and tracking all objects that are seen by the sensors. This algorithm is shown in Figure 10.

The clustering and tracking sensor fusion algorithm first uses information about the position and orientation of the sensors with respect to the vehicle to transform the returns into a vehicle centered coordinate system. Once the returns from the LIDARs are in vehicle centered coordinates, the position and orientation of the vehicle with respect to the world are used to transform the LIDAR returns into world coordinates. After the LIDAR returns have been transformed into world coordinates, they are clustered into groups of points. The clustering algorithm places the returns into a disjoint set data structure using a union find algorithm. Ultimately, clusters of returns are found whose members are not further than some maximum distance from each other. Once the LIDAR returns have been clustered, the resulting clusters must be tracked using dynamic filters. Vehicle detections that are returned by the vision system or the radar sensors are matched to a LIDAR generated cluster by looking for a LIDAR cluster within some distance threshold. If no suitable matching cluster is found, the detections may update or initialize a track without a corresponding LIDAR cluster. The output of the sensor fusion algorithm is a list of tracks. Each of the resulting tracks has a position and velocity, and the general size and shape of the point cluster supported the track is abstracted as a set of linear features.

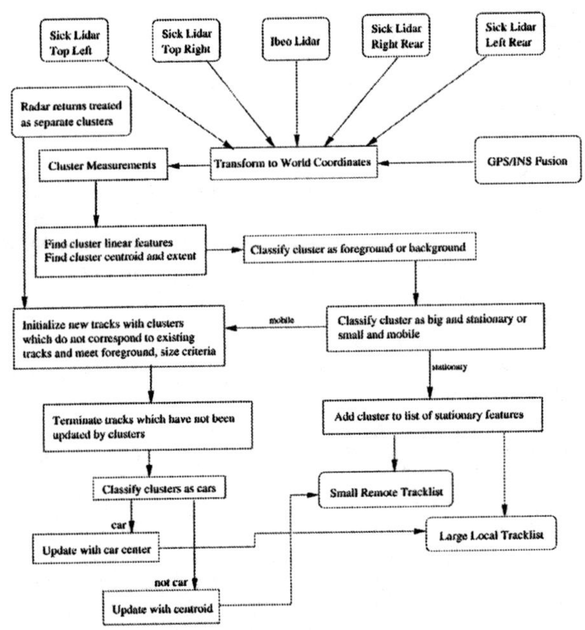

Figure 10: Block Diagram of OSU-ACT Sensor Fusion

The output of the clustering and tracking algorithm can be converted into a grid map if that is more convenient for a particular task, for example navigating an dense obstacle field such as might be found in a parking lot.

SENSOR INTERPRETATION AND SITUATION ANALYSIS

The term *situation* is defined to be knowledge concerning the vehicle and/or the prevailing scenario and surroundings. From a practical viewpoint, situations are the switching conditions among meta-states and all the sub-states inside the high-level control state-machines. Thus, the aim of situation analysis is to provide the high level controller with all the switching conditions in a timely manner. The situation analysis software analyzes the current vehicle state, the current and upcoming required behavior for the route plan, the map database, and the sensor data to identify specific situations and conditions that are relevant to the vehicle's immediate and planned behavior.

For the off-road Grand Challenge scenario, the situation is always obstacle and collision avoidance. The software is required to analyze the occupancy grid map, identify obstacles that may block the current path, and adjust the planned path as needed.

For an urban scenario, we are interested in all the targets in our path and the targets in surrounding lanes or on roads intersecting our lane. We are not interested in targets that do not affect the current situation and planned behaviour. While an autonomous vehicle is navigating through the city, many different situations may arise. The situations may vary if the vehicle is on a one-lane road, on a two-lane road, at an intersection, and so on. Particularly critical for an autonomous vehicle are those situations related to intersections. When a car is approaching an intersection, it must give precedence to other vehicles already stopped. If the intersection is not a four-way stop, the vehicle must cross or merge safely in the presence of oncoming traffic. If other vehicles are stationary for a long time, the car must decide whether those vehicles are showing indecisive behaviour. Other situations may involve road blockage in which the vehicle might carefully perform a U-turn, parking in parking spaces, and dealing with dangerous behaviour from other vehicles. All these situations must be identified and evaluated, and the resulting conclusions transmitted to the high level controller, in order for the vehicle to operate properly.

To allow situation analysis and sensor interpretation the OSU-ACT sensor fusion system provides information about the centroid's position and velocity as well as geometric shape information for each tracked cluster in a local navigation reference frame. The path planning software provides information related to the current optimal path plan. Starting from the path, the situation analyzer can identify the location of the road and build a road model constructed from polygons derived from a spline curve fitting the waypoints defining the road and lane shapes. Such a road model design is particularly suitable for both accuracy and implementation purposes. Using such a road model, traffic vehicles and other obstacles can be located and classified according to their lane positions and their impact on the current behaviour and plan for the autonomous vehicle's motion can be analyzed. In order to reduce computational costs and complexity, only the situations related to the current meta-state or sub-states, as provided by the high-level control software, are checked.

LOW LEVEL CONTROL

COMMAND INTERFACE - In a two-level control hierarchy as shown in Figure 4, the low-level control receives operational instructions from the high-level control module. These instructions take the form of

1. A path to be followed, defined by a set of approximately evenly spaced control points
2. A desired speed
3. Commands to indicate starting and stopping
4. Special commands indicating motions that can be fully implemented at the lower level. Examples could be short movements along constant radius arcs, precision stops, etc.

The low level control will execute a given command set until either the command is completed and the vehicle is in a stationary state, or until the vehicle has driven off the end of the path provided, at which point the vehicle will be stopped, or until it receives a new command set.

LONGITUDINAL CONTROL - The interface and control of vehicle actuation is achieved by having a drive-by-wire throttle and brake systems. Our experience has been that a simple control algorithm, for example a set of PID controllers, is adequate to generate a virtual torque command to achieve the commanded speed, and a state machine is used to select between the use of throttle, active braking, or engine idle braking. Speed commands are modified to constrain the acceleration and jerk of the vehicle to preset comfortable limits. There may also be "emergency" deceleration modes that are less comfortable.

Urban driving, in contrast to highway or off-road driving, requires the vehicle to also execute precise stops at predefined locations, for example the stop line of an intersection. To accomplish this, the low level control determines the distance from the vehicle's current position to a line drawn through the specified stopping point and perpendicular to the vehicle's path of travel, taking into consideration the distance from the front bumper of the vehicle to its centroid. The speed of the vehicle is controlled to follow a specified, possibly nonlinear, deceleration trajectory.

LATERAL CONTROL - The path that the vehicle is to follow is specified as a set of evenly spaced control points. The lateral controller identifies both the current location of the vehicle and the look-ahead point a pre-specified distance ahead of the vehicle along its lateral axis and extracts a subset of control points closest to each location. Constant radius circles are fitted to the points in each subset and these circles are used to compute the vehicle offset distances from the path and to estimate a desired yaw rates. Each subset of points also defines a desired yaw angle for the vehicle. The offset distances, yaw angle error measurements, and desired yaw rates can be used to generate a feedback signal for the steering controller. There are a number of algorithms that can be used in this control loop, and a simple PID controller with fixed gains is insufficient to cover all possible driving speeds and path-shape scenarios. The variations here are speed dependent and turn-radius dependent. See [12, 13] for some examples.

CONCLUSION

In this paper we have presented the autonomous vehicle systems used in our off-road DARPA Grand Challenge and DARPA Urban Challenge autonomous vehicles. We have described an architecture that is common to all autonomous vehicles. Some technological aspects are also common, such as vehicle localization, low level vehicle control, and path following. However, several technologies, including sensing, sensor fusion and interpretation, path planning, and behavior generation and high level control must be designed to match the level of autonomy and the scope of situations and behaviors required of the specific vehicle and task.

A key aspect of the successful implementation of an autonomous vehicle are the availability of adequate sensing technologies, which at this time may be quite expensive, and the ability and resources to perform extensive testing and evaluation in multiple and varied environments.

ACKNOWLEDGMENTS

The authors would like to thank the Ohio State University College of Engineering, National Instruments, ST Microelectronics, Oshkosh Truck Corporation, Topcon, and The Transportation Research Center (East Liberty, OH) for supporting these efforts. Projects of this scope reflect the work and contributions of many individuals; too many to be named here. Please refer to our websites [6,7].

REFERENCES

1. Defense Advanced Research Projects Agency, *Grand Challenge,* [online]. Available from http://www.darpa.mil/grandchallenge
2. Keith Redmill, Umit Ozguner, and Alberto Broggi, *Team TerraMax and the DARPA Grand Challenge: A General Overview,* Proceedings of the 2004 IEEE Intelligent Vehicles Symposium, 14-17 June 2004, pp. 232-237.
3. Qi Chen, Umit Ozguner, and Keith Redmill, *Ohio State University at the 2004 DARPA Grand Challenge: Developing a Completely Autonomous Vehicle*, IEEE Intelligent Systems Magazine, 19:5 (Sep-Oct 2004), pp. 8-11.
4. Qi Chen and Umit Ozguner, *Real-time Navigation for Autonomous Vehicles: a Fuzzy Obstacle Avoidance and Goal Approach Algorithm*, Proceedings of the 2005 American Control Conference, 8-10 June 2005, pp. 2153-2158.
5. Zhiyu Xiang and Umit Ozguner, *Environmental Perception and Multi-Sensor Data Fusion for Off-road Autonomous Vehicles*, Proceedings of the 2005 IEEE Intelligent Transportation Systems Conference, 13-16 Sept. 2005, pp. 584-589.
6. The Ohio State University, *Desert Buckeyes ION: The Intelligent Off-road Navigator*, [online]. Available from http://www.ece.osu.edu/ion.
7. The Ohio State University, *OSU-ACT The Autonomous City Transporter: 2007 DARPA Urban Challenge*, [online]. Available from http://www.ece.osu.edu/osuact.
8. Qi Chen and Umit Ozguner, *Intelligent Off-road Navigation Algorithms and Strategies of Team Desert Buckeyes in the DARPA Grand Challenge 2005*, Journal of Field Robotics, 23:9 (Sept 2006), pp. 729-743.
9. Keith A. Redmill, Takeshi Kitajima, and Umit Ozguner, *DGSP/INS Integrated Positioning for Control of Automated Vehicles*, Proceedings of the 2001 IEEE Intelligent Transportation Systems Conference, 25-29 August 2001, pp. 172-178.
10. Arda Kurt and Umit Ozguner, *Hybrid State System Development for Autonomous Vehicle Control in Urban Scenarios*, 2008 Proceedings of the IFAC World Congress, July 2008.
11. K. Redmill, J. I. Martin, and U. Ozguner, *Sensing and Sensor Fusion for the 2005 Desert Buckeyes DARPA Grand Challenge Offroad Autonomous Vehicle*, Proceedings of the 2006 IEEE Intelligent Vehicles Symposium, 13-15 June 2006, pp. 528-533.
12. K. A. "Unyelioglu, C. Hatipoglu, and U. Ozguner, *Design and Stability Analysis Of A Lane Following Controller*, IEEE Transactions on Control Systems Technology, 5:1 (January 1997), pp. 127-134.
13. C. Hatipoglu, U. Ozguner, and K. A. Redmill, *Automated Lane Change Controller Design for Intelligent Transportation Systems*, IEEE Transactions on Intelligent Transportation Systems, 4:1 (March 2003), pp. 13-22.

CONTACT

Keith Redmill (redmill@ece.osu.edu) received the B.S.E.E. and B.A in Mathematics from Duke University, Durham, NC in 1989 and the M.S. and Ph.D. degrees from The Ohio State University, Columbus in 1991 and 1998 respectively. He has been a Research Scientist in the Department of Electrical and Computer Engineering since 1998.

2008-01-2717

Low-Cost Autonomous Vehicles for Urban Environments

Mahesh K. Chengalva and Richard Bletsis
Delphi E&S

Bernard P. Moss
Control-Point Corp.

ABSTRACT

Despite the rapid progress in the development of autonomous vehicles, as seen in the DARPA Urban Challenge 2007, there has been very little emphasis on minimizing costs. Some teams spent upward of $10 million in developmental expenses. The cost factor is very important as it is the primary driver leading to the commercialization of autonomous technology. With this fact in mind, an alternative approach has been emphasized here, wherein a fully autonomous vehicle designed for urban environments has been developed and tested for under $20,000 in hardware costs. Moreover, this vehicle passed several rounds of elimination to participate in the semi-finals of the DARPA Urban Challenge at Victorville, California in November 2007.

OVERVIEW

While there has been considerable development in the area of autonomous vehicles in recent years, the emphasis on low cost solutions has been neglected. As seen in the DARPA Urban Challenge held in November 2007, some of the teams had spent in excess of ten million dollars on their vehicles. Commercialization of autonomous vehicle technology however demands that cost issues remain at the forefront. This paper describes an autonomous vehicle system with less than $20,000 of hardware on board that has been extensively field-tested and which participated in the concluding rounds of the recent DARPA Urban Challenge.

The primary drivers of a low-cost solution to autonomous urban navigation are the use of low-cost commercial hardware along with their interfacing with software containing sophisticated algorithms able to maximize the utilization of the associated hardware. These algorithms are intensively mathematical, often employing closed-form solutions, to not only interface with relatively simple sensors but also to simplify computations so that a single desktop PC is sufficient to handle all aspects of vehicle control.

One example of this pairing of simple hardware and sophisticated software is in the vision system, which is the primary sensor system developed for our autonomous vehicle. The vision system hardware utilizes several $50 webcams that provide all the necessary input. To reduce computational requirements, all cameras operate in the 320 x 240 pixel resolution only. Vision algorithms were created to simultaneously process multiple camera data and utilize these to obtain a real-time 3-D visualization of the surrounding spatial environment. Moreover, the computational efficiency of these algorithms permitted the use of a single desktop PC along with the flexibility of using Visual Basic 6 as the programming language.

Another such example is in the brake and throttle control where simple linear actuators are connected to the transmission shifter, accelerator and brake pedals of the vehicle, and are integrated in a manner that provides fail-safe operation via mechanical means. In the event that pre-set driving parameters are exceeded, or any type of power failure occurs, the vehicle will always come to a quick stop. These actuators are controlled by special algorithms employing closed-form equations to provide real-time dynamic control of the vehicle based on just the actual and desired velocities as input. Another critical innovation employed in the vehicle is the creation of a virtual Inertial Measurement Unit (IMU) which utilizes existing vehicle sensors and special algorithms, thereby eliminating the need for purchasing very expensive IMU hardware.

The research and developmental activity that preceded the testing of the full-size autonomous vehicle also features many innovative low-cost solutions. Chief among these is the extensive use of scale models to develop critical vision algorithms that form the core of the vehicle control system. Initial development occurred with 1:48 scale static models. Algorithms that passed this stage were encoded into a specially fabricated 1:3 scale self-propelled autonomous vehicle and tested for dynamic validation with other 1:3 scale traffic vehicles. Only when the algorithms passed these two stages were they tested on the full-size vehicle, which not only saved considerably on the cost of testing, but also significantly lowered the risk factor for damage or injury at the full-size vehicle level. Despite the inherently high risk of testing a newly developed autonomous vehicle, no person or property suffered even a minor scratch throughout this endeavor that lasted well over one year. This can be attributed to strict adherence to proper safety procedures by all participants, and the innovative fail-safe designs that were implemented.

There are two other significant advantages of our system. The first of these is that it is ideal for military applications because it utilizes a passive vision system as the primary sensor system. Unlike other vehicles that employ a number of radars which greatly increases their electromagnetic signature, our vehicle is not only passive but, due to the small size of the cameras, presents a very low physical profile. The other advantage of our system is in the ease with which it can be converted from fully autonomous to fully manual control. This can be accomplished in seconds because the special design of the actuator assembly enables manual override without the need for any removal of actuator hardware. A safety driver during autonomous testing (which was always performed in approved off-road locations), could take over control of the vehicle at any instant, immediately – without the need of pushing any switches or levers to transition from autonomous to human operation.

In summary, a versatile low-cost autonomous vehicle system suited for navigation in urban environments is presented here, which has been extensively field-tested and which also successfully participated in the DARPA Urban Challenge in 2007.

DEFINITIONS

Ariadne
Team ODY-ERA's position-based, precision Ded Reckoning system.

Cyclops
A novel machine vision system invented and developed by Team ODY-ERA for use in the Urban Challenge and other future applications. Appropriately named after the primordial race of one-eyed giants from Greek mythology, the ODY-ERA vision system primarily uses only a single forward-looking camera.

CVT
Continuously Variable Transmission

GPS
Global Positioning System

Hanuman
The integrated system of sensors and software that controls our autonomous vehicle, named after the legendary warrior Hanuman from the ancient Indian epic - the Ramayana. Known for his immense strength, unflinching loyalty and the ability to effortlessly travel to any place at any time and under any conditions, Hanuman embodies the very qualities we seek in our autonomous vehicle.

IMU
Inertial Measurement Unit

IR
Infra-Red

LMS
Laser Measuring System (also known as LIDAR)

MDF
Mission Data File, a file format defined by DARPA for use in the Urban Challenge.

ODY-ERA
ODYsseus-ERAtosthenes — An epic combination of elegant and efficient mechanical, navigational, numerical, and algorithmic concepts, implemented in the creation of an autonomous ground vehicle. Odysseus and Eratosthenes represent two giant figures from Antiquity on whose shoulders we are standing.

RC
Remote Control

RNDF
Route Network Definition File, a file format defined by DARPA for use in the Urban Challenge.

SUV
Sport Utility Vehicle

USB
Universal Serial Bus

HARDWARE

After careful deliberation, a vehicle platform that was considered to be best suited for the Urban Challenge 2007 competition – a new 2008 Mercury Mariner Hybrid 4WD SUV (shown in Figure 1) was selected. The first and most important consideration is that this vehicle meets DARPA Urban Challenge specifications. The combination of an electric motor and a gasoline engine for propulsion coupled to a CVT results in smooth operation and requires minimal modifications. Its electrical system can be utilized to draw up to 1200W to power the ODY-ERA computer and autonomous vehicle control systems. Furthermore, the engine will cycle on and off automatically while on "Pause" during a mission, as needed to maintain a controlled climate in the cabin and proper levels of electrical power to the equipment.

Figure 1: The 2008 Mercury Mariner Hybrid 4WD, ODY-ERA autonomous vehicle.

Even though computers and software are of utmost importance in the development of this project, the mechanical aspects have also been given thorough consideration. As a matter of fact, it is the synergistic effects from several proven mechanical innovations we have already developed, that enabled us to bring an elegantly simple autonomous solution to this competition. Using our experiences from building a custom robotic vehicle, called "Conestoga-Bot", which advanced in the quarterfinals of the Grand Challenge 2005 (for the most part due to the merits of its physical capabilities), we implemented several vehicle control features we had already developed to the Mercury Mariner Hybrid SUV, as part of its transformation into ODY-ERA. One of these is our precision closed-loop, servo-controlled steering system we have implemented, while retaining all stock electric-steering components of the Mercury Mariner. Three other innovative features we had designed for the Conestoga-Bot and reused on ODY-ERA are the unobtrusive gear-selector, throttle, and brake control sub-systems that provide fail-safe autonomous operation when used with our E-stop, or the E-stop that is provided by DARPA. Furthermore, these systems allow for immediate, unencumbered transition from manual to autonomous vehicle operation and vice versa. During testing and development, a safety person sitting in the driver's seat can instantly take control of the vehicle at any time, as needed. All of the vehicle's original safety restraints and airbags remain intact and offer security to the occupant(s) while testing, as was intended by the manufacturer for regular road use.

The largest problem associated with this program is the inherently dangerous nature of having a 2-ton vehicle operating autonomously during its development. Since safety is the number one priority of Team ODY-ERA, considerable effort has been directed toward mitigating all circumstances that can pose a threat to the personal well-being of crew and others and/or to property. While our meticulous approach has resulted in affording ample security to persons in the vicinity and to the occupant(s) of the ODY-ERA vehicle while testing, our initial development of algorithms and software are previously validated with small-scale models. For example, vision algorithms are at first developed using still pictures. Then a 1:48 scale model of roadway, intersection and Matchbox-style cars are used along with a computer and video cameras to confirm the desired results. Once satisfactory results are obtained, the next step is to conduct autonomous testing with "mOE" (mini ODY-ERA) — a 1:3 scale vehicle we have specifically designed and built to reduce exposure and thus enhance safety, as seen in Figures 2 and 3.

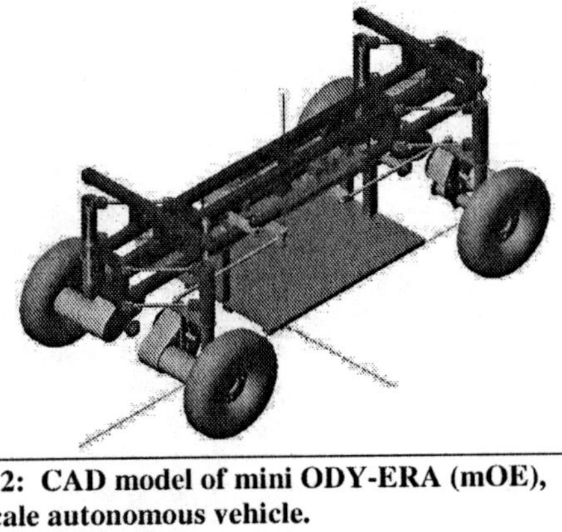

Figure 2: CAD model of mini ODY-ERA (mOE), a 1:3 scale autonomous vehicle.

This vehicle is powered by two 12Vdc utility vehicle batteries that drive four DC motors, one at each wheel. Steering and propulsion are controlled by a RoboteQ AX2850. Commands for vehicle speed and steering are communicated to the RoboteQ via RS232 from a WindowsXP-based laptop PC. The autonomous software code processes RNDF and MDF files, along with data from GPS, E-compass, video cameras, and other sensors. In addition to the benefits gained in terms of increased safety of our program by small-scale testing, we have proven that our approach in developing an autonomous ground vehicle system is scaleable and can be readily adapted to any size vehicle.

Figure 3: Mini ODY-ERA (mOE), custom-made 1:3 scale autonomous vehicle.

As a matter of fact, we naturally find that if a certain version of the code works satisfactorily with mOE, we end up with better, more precise results when we scale up to run the full-size ODY-ERA vehicle. This is due to the fact that the GPS uncertainty is the same whether you are operating on a (1:3 scale) 5ft wide lane or a (full scale) 15ft wide lane.

MECHANICAL EVALUATIONS

Prior to installing any of the systems for autonomous operation to the Mercury Mariner SUV, vehicle testing was conducted in large vacant paved and unpaved areas, such as parking lots and farms, with level and inclined terrain. Extensive data was gathered by using accelerometers, where the performance envelope of the vehicle was determined in terms of throttle, braking, and cornering response for speeds of up to 30 mph. Data was also collected on wheel position with respect to distance traveled while going straight and at several different steering angles. It was then reduced into tabular form and closed-form equations for more efficient programming. This information was later used for calibration and to define competitive, but safe limits of autonomous vehicle operation. As an example, maximum vehicle speed was defined with respect to steering angle for various conditions, as shown in Figure 4. A threshold limit curve is defined for smooth and level paved road, as shown, which will never be exceeded – even though it includes a generous factor of safety.

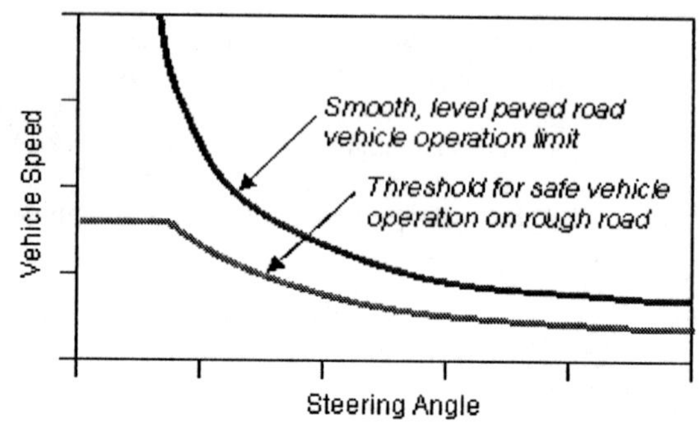

Figure 4: Speed-versus-steering performance envelope.

Similar curves were defined for increasingly worse conditions and have also been included in the algorithms for further de-rating (limiting) allowable combinations of vehicle speeds/steering angles. This performance envelope of the vehicle is used extensively in combination with routing algorithms for predicting the target speed of the vehicle throughout a mission.

ROUTING ALGORITHMS

Determination of the most optimum route within the RNDF for any given mission is accomplished by a specially developed variation of Dijkstra's Algorithm [1] with additional considerations embedded. All potential routes are considered between two checkpoints, and then weighted penalty-functions are used to calculate the quickest route. Distinct penalty-functions are assigned for crossing 4-way-stop intersections, merging into thru traffic, speed limits, turns at intersections, etc. By considering all these factors in the routing calculations, oftentimes the shortest route is not the quickest. Once a roadblock is encountered, the blocked section can be avoided from either side for the remainder of a mission.

A typical urban route that is about 10 miles long is calculated in a fraction of a second, along with creating a Ded Reckoning database that includes heading and true position at 1ft intervals with target vehicle speed, target steering angle, and elapsed time (ET) at each interval. The Ded Reckoning predictions are based on vehicle calibration, performance envelope, calculated road curvature, and penalty functions – assuming smooth and level paved road. This information is available in real-time and is used in conjunction with the vision system. The vision system, which is also supplemented by other sensors when applicable, considers the target speeds as its speed limits and de-rates them as necessary based on conditions. Sometimes, when vision does not adequately detect road roughness, input from the accelerometers is used to further de-rate vehicle speed. A similar way of de-rating speed and modifying steering angle is done when other supplemental sensors provide conflicting information, based on a decision matrix that utilizes and solves such Sorites-style arguments [2]. Known deviations from the initial predictions are continually used to update the Ded Reckoning database.

When an unexpected roadblock is encountered, ODY-ERA executes a 3-point (180° CCW) turn, while the route to the remaining checkpoints is recalculated, and the Ded Reckoning database is updated. The typical 3-point turn takes less than two minutes to execute. During this time, all of the re-routing calculations have been made and there is no need for further delay.

The low budget of ODY-ERA has forced the team to innovate and come up with straight-forward, simple and inexpensive solutions. Instead of an unaffordable high-end IMU, we have developed "Ariadne"; a low-cost, precision Ded Reckoning system. By not being able to afford 3D LMS, we opted to invent and develop a more effective solution of machine vision (Cyclops), which is extremely accurate and can be readily implemented for the most part with low-cost USB cameras. All the low-cost innovations are assembled into a robust autonomous ground vehicle system, code-named Hanuman.

AUTONOMOUS SYSTEM ARCHITECTURE

The ODY-ERA mechatronic system was designed to be powered by the vehicle's existing electrical configuration. The main objectives of the design were safety (E-stop conformance, easy transition from autonomous to manual control of driving), low power consumption and 12Vdc power source. Figure 5 shows a block diagram of the ODY-ERA architecture.

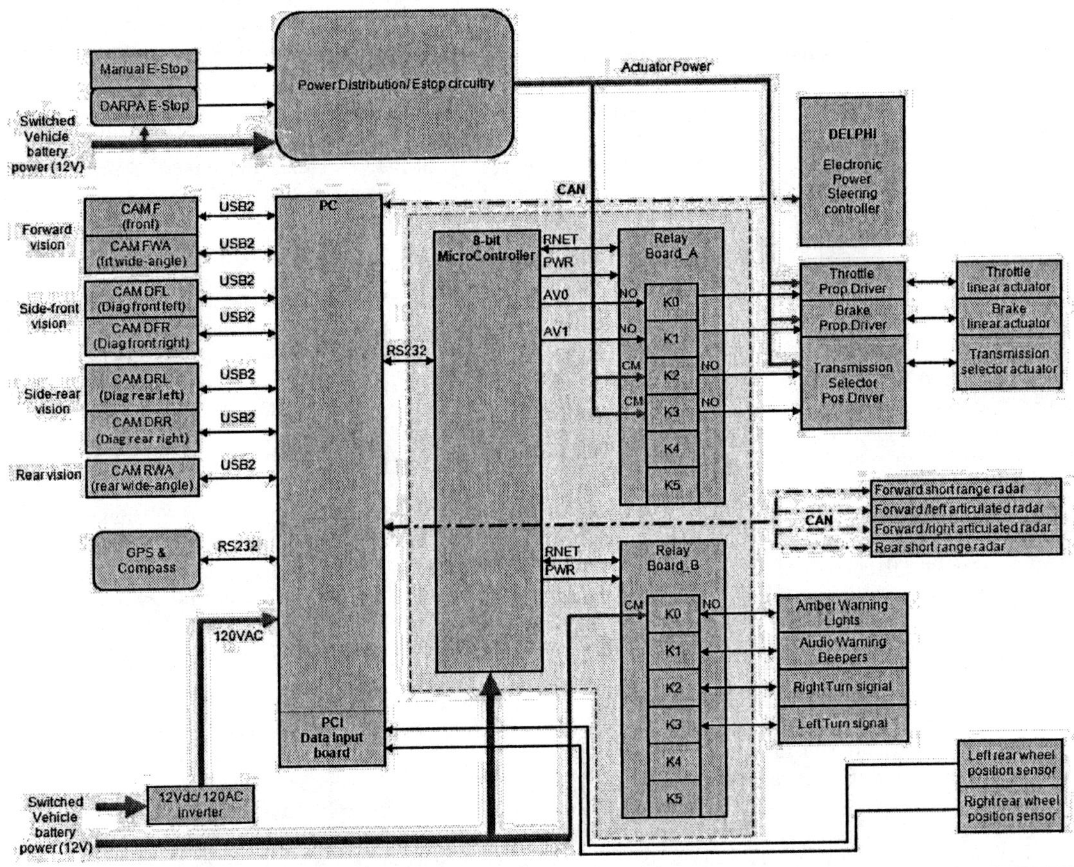

Figure 5: ODY-ERA Block Diagram of Autonomous System Architecture.

The system design is fail-safe and guarantees that E-stop will always bring the vehicle to a controlled stop – even if the ODY-ERA computer stops functioning, or if power is lost. The "Pause" function is implemented via the computer and software. Furthermore, the team-supplied E-stop receiver constantly queries the transmitter for a coded signal and if the vehicle is out of range, power to the throttle and brake controllers is interrupted, which results in clutch disengagement at the actuators and fail-safe mechanical application of the brakes to promptly halt the vehicle. The system allows for a quick and easy (on-demand) switchover from autonomous to manual control without necessitating hardware changeovers. During all autonomous testing, a person sitting in the driver's seat of the SUV can immediately take over control of the vehicle should the circumstances warrant it. In order to obviate the need for an auxiliary power system, ODY-ERA components have been selected so that the overall power consumption is minimized.

VISION SYSTEM OVERVIEW

A highly unique vision system has been developed as the primary sensor system for autonomous navigation, as will be described in this section.

From the military perspective, electromagnetic emissions from a vehicle should be avoided as much as possible, due to risk of vehicle detection in a hostile environment. This fundamental fact played a major role in determining that passive vision sensing should be the primary sensor system configuration for the ODY-ERA vehicle.

Before going into any more details of the vision system, it should be noted that the current vision system can readily be adapted to the use of infrared imaging cameras. The algorithms used for vision processing are sufficiently generalized to make this transition with minimal changes. More details on these algorithms are provided later in this section.

The vision system for the ODY-ERA vehicle consists of a suite of video cameras, mounted as sketched in Figure 6. Basically, this consists of two forward looking cameras, one of which has a "fish-eye" lens to capture a very wide field of view. Two additional cameras are present near the front, which are offset from the straight-ahead position by 50 degrees to the left and right. Slightly aft of these, two additional cameras are oriented about 110 degrees from the straight-ahead position. Lastly, a video camera with a fish-eye lens is pointed directly behind to enable rear vision capability.

These seven cameras enable the ODY-ERA vehicle to have an almost 360 degree sweep of visibility, with only a few small blind spots. Each camera has a field of view of 50 degrees with the exception of the cameras with the fish-eye lens which is capable of a 150 degree field of view.

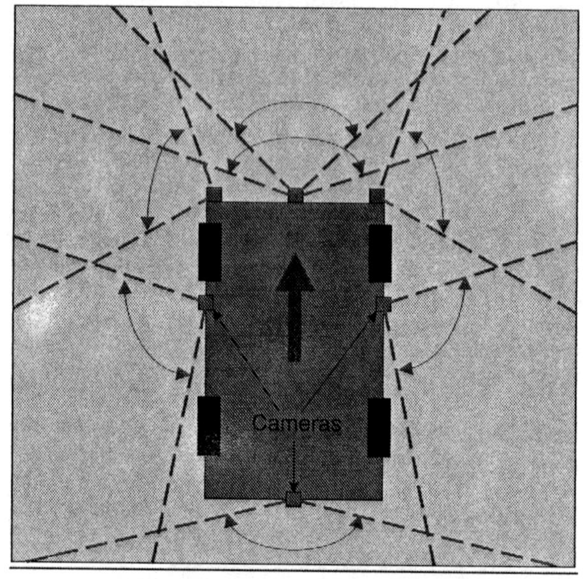

Figure 6: Schematic showing the layout of cameras with overlapping field-of-views to ensure vision coverage around most of the vehicle. The location at the front center of the vehicle consists of both a regular as well as a 'fish-eye' lens camera.

All cameras operate at either at 640x480 or at 320x240 resolution. The input consists of three color values for each pixel, one for each of the primary colors, in the 0 to 255 range for each color. This gives rise to a total number of color combinations of over 16 million (255 x 255 x 255) unique individual colors for each pixel. The total number of pixels per frame is either around 300,000 for the 640x480 video image or around 76,000 for the 320x240 image. This information is updated between 10 and 30 times per second from each camera. As can be seen from the above data, with all seven cameras simultaneously operating, the amount of information streaming into the vision processing unit can be substantial. All these data need to be processed by a single desktop computer. This is achieved by special screening algorithms that selectively process only the relevant information at any given instant, which is typically only a small fraction of the total information.

VISION PROCESSING ALGORITHMS

The data streaming in real-time from the various on-board video-cameras is processed using sophisticated algorithms to interpret, analyze and control the vehicle. These algorithms form the core of the vision system and will be briefly described in the following paragraphs.

One of the fundamental differences between the ODY-ERA vision system and other systems available today is that in the ODY-ERA system, the emphasis is on determining the mathematical characteristics of the detected object, rather than its physical attributes. For example, the ODY-ERA vision system does not care if the object blocking the road ahead is a set of traffic cones or a tank. All it cares about is the extents of the obstacle, its position relative to the ODY-ERA vehicle and its velocity vector.

From the algorithm perspective, there is considerable advantage to be gained from this approach. Limited processing resources can be utilized much more optimally. A greatly simplified algorithm structure is another advantage, along with the inherent increase in reliability due to simplification of the overall system complexity.

It should be noted that the above approach has been successfully employed in the ODY-ERA vehicle, as has been demonstrated by the almost 1000 miles of autonomous travel during numerous developmental runs as well as those at the DARPA Urban Challenge.

The governing algorithms of the vision system are highly mathematical in nature. The fundamental basis of these algorithms is the existence of an imaginary field in front of the vehicle, which is disturbed by the presence of obstacles. For the purpose of basic visualization, this can be compared to an electromagnetic or gravitational field, but this simplification should not be generalized as there are significant differences.

This vector field simultaneously exists in two spaces, with a one-to-one translation between individual locations in these two spaces. The first space in which this field operates is that of the images generated by the individual video cameras and defined by the axes U and V. This is shown in Figure 7.

The second space is the physical space in which the vehicle operates. This is illustrated in Figure 8. Note the use of the Eulerian coordinate system, wherein the coordinate system origin is attached to the vehicle.

Based on Figures 7 and 8, the field equations can be written as:

$$\mathbf{F}_a = \mathbf{F}_a(u, v, t) \tag{1}$$

and

$$\mathbf{F}_b = \mathbf{F}_b(x, y, z, t) \tag{2}$$

Where u, v are coordinates in the image frame, as shown in Figure 2, while x, y, z are coordinates in the vehicle reference frame shown in Figure 3. In both equations, 't' corresponds to time.

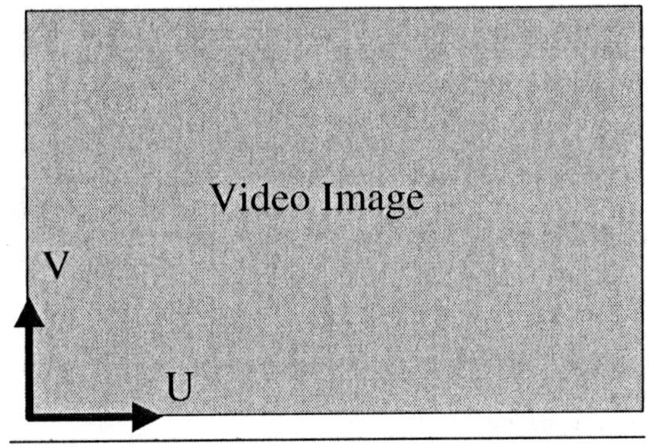

Figure 7: Coordinate axes defining the space in which a 'field' is generated.

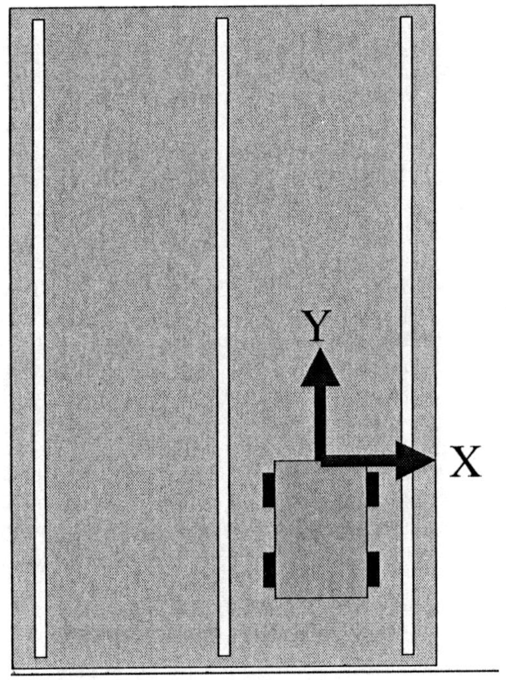

Figure 8: Coordinate axes defining the physical space around the autonomous vehicle. The Z axis is out of the plane of the figure.

The one-to-one mapping between the u-v and the x-y-z spaces can be represented by the matrix equation:

$$\mathbf{X} = \mathbf{M} * \mathbf{U} \quad (3)$$

where X and U are vectors consisting of components (x, y, z) and (u, v) respectively, and M is a transformation matrix. More information on the mathematical aspects of this section is available in references [3] and [4].

In the total absence of any obstacles or objects in the field of view, the field $\mathbf{F}_a$ assumes a constant vector everywhere. However, the presence of objects distorts this field in the same manner as an electric charge distorts an electric field.

In the first step of the computations, the field $\mathbf{F}_a$ is determined utilizing detailed pixel color information as mapped in Red-Green-Blue (RGB) space. This three-dimensional RGB space conveniently permits the utilization of vector calculus in determining the time-dependent field $\mathbf{F}_a$.

Once $\mathbf{F}_a$ is determined, potential obstacles are identified by the spatial gradients produced by their presence in the field. High gradients correspond to potential obstacles when certain other conditions are satisfied, which are then marked in u-v space by means of 'dots', which are essentially dimensionless entities. These dots are then transferred to x-y-z space by means of the transformation matrix **M**, as described in Equation (3).

While this approach may appear similar to other contrast-driven algorithms for obstacle detection, the utilization of field theory coupled with the flexibility of vector calculus overcomes many limitations that would otherwise be encountered. For example, shadows of objects result in high contrast regions in the image which can potentially confuse a vision system that relies purely on contrasts. This is greatly minimized in our vision system by considering the mathematical characteristics of shadow fields and appropriately correcting the image prior to processing.

The results of the above transformation provide a spatial distribution of obstacles around the vehicle. This distribution is refreshed between 10 and 30 times per second, providing the autonomous navigation system with an extensive real-time situational awareness.

As can be inferred from the description, there are two key challenges in the above approach. The first of these is the derivation of the spatial field $\mathbf{F}_a$ and the second is the computation of the transformation matrix **M**. Due to the intensive technical nature of these two derivations, the detailed mathematical descriptions of these derivations far exceed the allowed length of this technical paper. It is currently intended that these derivations to be disclosed in technical papers at a later date.

IMPLEMENTATION OF THE VISION ALGORITHMS

The algorithms described in the earlier section have been implemented and tested on various scales. A 1:48 scale model of a mock intersection was created which utilizes scale vehicles for added realism for testing the vision system (Figure 9). Note the presence of the 'vehicle' with a camera. In addition, a 1:3 scale prototype was fabricated and tests conducted with full autonomous navigation capabilities (Figure 3).

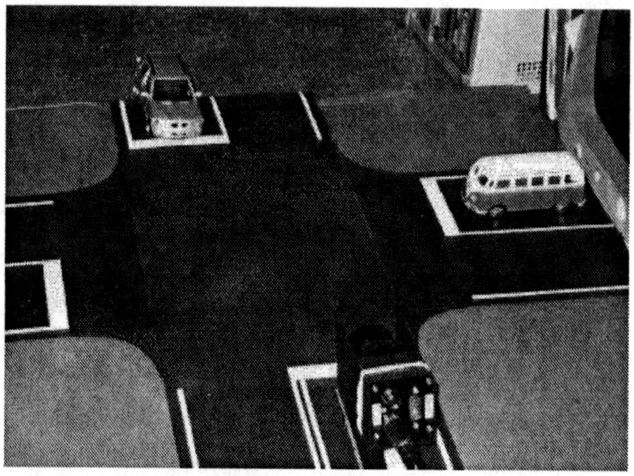

Figure 9: The 1:48 scale model of an intersection being utilized for algorithm verification.

A full scale vehicle has also been extensively tested in autonomous operation, along with participation in the DARPA Urban Challenge. All three scales of operation utilize exactly the same algorithms in the software so as to enable translation of results from tests conducted at one scale to another. Interestingly, the same cameras have been utilized at all three scaling configurations, including the 1:48 configuration.

The algorithms as described in the previous section can now be illustrated using the 1:3 scaled prototype.

Figure 10 shows a frame extracted from the video input of the 1:3 scaled prototype vehicle, representing a typical roadway encountered by the vision system during testing. Note that this roadway does not have any lane markings, thereby making it harder to analyze than one that has clear lane markings.

The information contained in this frame is analyzed by the vision algorithms to construct the field defined by Equation (1). After initial analysis, spatial gradients of this field are then computed. This is further processed to provide a set of 'dots' indicating limits of the roadway, as well as detecting any potential obstacles. Figure 11 shows the output from this stage of processing.

Figure 10: Frame extracted from the video camera output for the 1:3 scale prototype, prior to any image processing.

Figure 11: Image processing to derive the field defined by Equation (1), followed by further processing to define the limits of the roadway.

The 'dots' shown in Figure 11 are projected utilizing Equation (3) for determining the extent of the roadway ahead of the vehicle. This is shown in Figure 12. Depth perception is obtained to a distance of 50 feet in this case, which corresponds to a distance of 150 feet in an actual full-scale vehicle (moving objects have been detected at greater ranges utilizing time derivatives of the field $\mathbf{F}_a$).

The resulting information provides a valuable map of obstacles in the immediate vicinity of the vehicle, based on which a desired trajectory (with corresponding desired velocities) is computed, which is then translated into steering and throttle/brake commands. The above process can be repeated as many as 10-30 times per second, enabling close tracking of other moving objects in the vicinity of the autonomous vehicle.

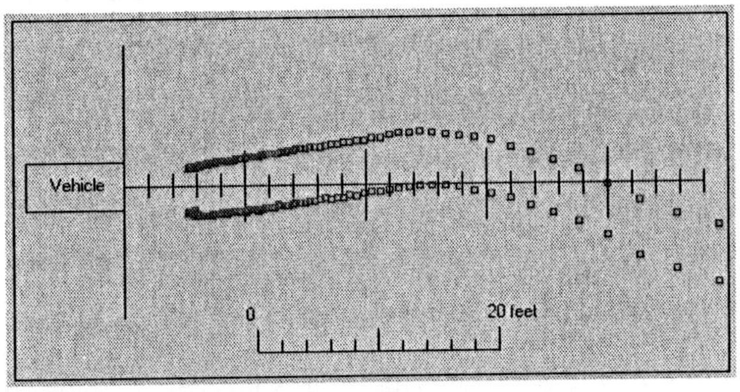

Figure 12: Projection of 'dots' utilizing Equation (3) to obtain spatial depth perception.

The same principle described above on the one-third scale vehicle is employed in the full-scale vehicle. A similar analysis is shown for an actual road environment. Figure 13 represents a typical traffic stop with a stationary vehicle ahead. Figure 14 schematically represents the spatial depth perception derived from the image. Two separate sets of algorithms operate simultaneously on the image. The first of these computes the desired trajectory by identifying lane markings on either side. These algorithms work even when no distinct lane marking exist, such as the example shown in Figure 10. The final output of these algorithms is the instantaneous steering angle, which is the angle by which the steering wheel needs to be turned. Since updates to the steering angle are made over ten times per second, very smooth and steady steering behavior is obtained.

Figure 13: A typical situation encountered in regular traffic.

The second set of algorithms identify obstacles and then determine an instantaneous desired velocity, which the control system ensures is followed as closely as possible. Once again, because desired velocity updates are made over ten times per second, the vehicle exhibits smooth braking and acceleration characteristics.

Two points of interest should be noted in Figure 14. Firstly, no attempt is made to identify the obstacle as a vehicle. As far as the vision system is concerned, the only attributes of the obstacle that matter are its size, position and relative velocity. Secondly, all the data in Figure 14 has been obtained by a single forward-facing camera. Depth perception is derived from Equation 3.

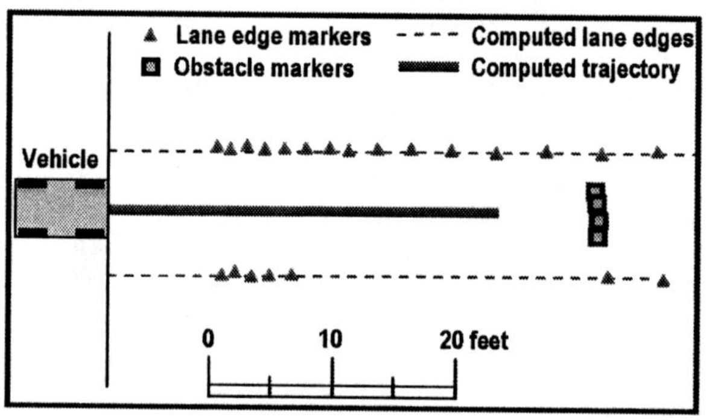

Figure 14: Results of processing image shown in Figure 13, showing desired vehicle trajectory.

The system is capable of reliably detecting obstacles as small as a tennis ball at close range. It has been tested with various obstacles on the roadway utilizing the 1:3 scale prototype vehicle, including traffic cones, other vehicles, barricades, etc. With additional processing utilizing the time dependency of Equation (1), it is even possible to determine approximately the velocities of other moving vehicles. This capability is utilized at intersections, such as a four-way stop, to determine priority of passage through the intersection.

However, this system can, on rare occasions, produce false positive results, i.e., the system detects an object that that should really be of minor or no concern. In such cases, alternative sensor systems on board, such as the radar or ultrasound sensors, can enable rapid clarification of the nature of the obstacle. It should be noted that the opposite case (not detecting objects of concern) has not occurred so far in all the extensive testing that has been conducted.

It should also be noted that although the associated theory may appear abstract, it has been successfully implemented in a full-size vehicle, which has been extensively tested. During these tests, it has been observed that the vision algorithms provide very smooth and robust control of the vehicle, while at the same time requiring only minimal computing power. It is mainly due to the computational efficiency of these algorithms that a single Windows-based desktop PC is the only computing hardware present in the vehicle.

This ties in with the cost factor: since most of the expertise resides in the software, which essentially encapsulates algorithms that are not only highly mathematical but also highly efficient, the hardware expenses are greatly minimized. For example, the total cost of the hardware on board the ODY-ERA vehicle, including the computer, all cameras, GPS, other sensors and actuators, was less than $20,000. In fact, side-vision sensors in the vehicle essentially use only $50 web cams as the only image input devices at a 320 x 240 resolution.

An additional advantage of the vision system is the scalability aspect, as demonstrated by the utilization of the entire system on a 1:48 scale, then on a 1:3 scale prototype, followed by the use on a full-scale vehicle. Almost no modifications were required other than changes to a few obvious parametric variables in order to obtain full functionality irrespective of scale. The utilization of the identical system on a 1:3 scale prototype as well as a full-sized vehicle clearly demonstrates the potential of this vision system in terms of transferability between vehicles of different sizes with minimal modifications.

SUMMARY

A low-cost autonomous vehicle is presented here that has been extensively field tested and which also passed several rounds of eliminations to participate as a semi-finalist in the DARPA Urban Challenge (photos in Figures 15, 16; videos in Reference [5]). As also reported widely in the media, the ODY-ERA vehicle is the lowest budget vehicle to qualify for the semi-final rounds of competition, employing only a fraction of the resources of other competing teams.

This has been only possible due to the numerous unique innovations employed, from the virtual Inertial Measurement Unit which utilizes data available in any stock vehicle to closed-form equations that provide smooth control of acceleration and braking without the need for accelerometers and load-cells.

A unique vision system is the backbone of the ODY-ERA vehicle. The low-cost, high-degree of adaptability and scalability as well as the inherent robustness of the algorithms that constitute the vision system makes this a highly promising system for autonomous vehicles. The emphasis on passive sensing and minimal vehicle profile disruptions by sensing hardware also makes this system ideal for autonomous military vehicles.

Figure 15: Conducting an autonomous mission in an urban setting, while safely sharing the road with over a dozen other vehicles at the DARPA Urban Challenge.

Figure 16: The ODY-ERA vehicle stops at an intersection to yield to oncoming traffic, prior to making a left turn.

REFERENCES

[1] "Dijkstra's algorithm", National Institute of Standards and Technology (NIST), http://www.nist.gov/dads/HTML/dijkstraalgo.html

[2] "Sorites paradox", Stanford Encyclopedia of Philosophy, http://plato.stanford.edu/entries/sorites-paradox/

[3] "Advanced Engineering Mathematics", Erwin Kreyszig, John Wiley & Sons, Inc., 1998.

[4] "Tensor Calculus", David C. Kay, McGraw-Hill, Inc., 1988.

[5] ODY-ERA Website: www.geocities.com/odyera2007

ACKNOWLEDGEMENTS

The authors would like to acknowledge the other ODY-ERA team members, all of whom have worked diligently and tirelessly to keep ODY-ERA competitive in the DARPA Urban Challenge. They are, alphabetically: Mac Gray, Stavros Kritikos and Ronald Webster.

Also acknowledged are extended team members Eric Miller and Gerry VanHorn, who helped the team extensively during the last two months of the Urban Challenge.

The ODY-ERA Team would like to thank its major sponsor, Delphi (www.delphi.com), for providing equipment, sensor integration support, and transporting the vehicles to and from the Urban Challenge site.

OmniVision www.ovt.com is acknowledged for donating cameras to ODY-ERA, and Control Point Corporation www.control-pt.com for expert electrical systems development.

SAE International

Vehicle Safety Communications - Applications: System Design & Objective Testing Results

2011-01-0575
Published
04/12/2011

Farid Ahmed-Zaid
Ford Motor Company

Hariharan Krishnan
General Motors Company

Michael Maile
Mercedes Benz REDNA

Lorenzo Caminiti
Toyota Motor Engineering & Mfg NA Inc.

Sue Bai
Honda R&D Americas Inc.

Steve VanSickle
Danlaw, Inc.

doi:10.4271/2011-01-0575

ABSTRACT

The USDOT and the Crash Avoidance Metrics Partnership-Vehicle Safety Communications 2 (CAMP-VSC2) Consortium (Ford, GM, Honda, Mercedes, and Toyota) initiated, in December 2006, a three-year collaborative effort in the area of wireless-based safety applications under the Vehicle Safety Communications-Applications (VSC-A) Project. The VSC-A Project developed and tested communications-based vehicle safety systems to determine if Dedicated Short Range Communications (DSRC) at 5.9 GHz, in combination with vehicle positioning, would improve upon autonomous vehicle-based safety systems and/or enable new communications-based safety applications. The project addressed the following objectives:

- Assess how previously identified crash-imminent safety scenarios in autonomous systems could be addressed and improved by DSRC+Positioning systems
- Define a set of DSRC+Positioning based vehicle safety applications and application specifications including minimum system performance requirements
- Develop scalable, common vehicle safety communication architecture, protocols, and messaging framework (interfaces) necessary to achieve interoperability and cohesiveness among different vehicle manufacturers. Standardize this messaging framework and the communication protocols (including message sets) to facilitate future deployment.
- Develop requirements for accurate and affordable vehicle positioning technology needed, in conjunction with the 5.9 GHz DSRC, to support most of the safety applications with high-potential benefits
- Develop and verify a set of objective test procedures for the vehicle safety communications applications

In this paper, we summarize the work that took place in the VSC-A Project in the areas of system design and objective testing. We first introduce the VSC-A system framework. We then list the crash imminent scenarios addressed by the VSC-A Project and the safety applications selected to potentially address them. Next we describe the VSC-A test bed system development. This test bed was ultimately used to verify Vehicle-to-Vehicle (V2V) communication interoperability between Ford, GM, Honda, Mercedes-Benz, and Toyota vehicles. Public demonstrations of V2V interoperability were held in New York City at the 2008 Intelligent Transport Systems (ITS) World Congress. The test bed also served to validate the system and minimum performance specifications that were developed as part of this project. We discuss one of the most important achievements of the project in the communication area, i.e., implementation, testing, verification, and standardization of a safety message that supports all of the VSC-A safety applications. The result is the Basic Safety Message (BSM) as defined in the SAE J2735 Message Set Dictionary standard. Details of the objective test procedures are presented next and are followed by a summary of the performed test scenarios (test descriptions, speeds, number of runs for each test, type of test, etc.) with the corresponding objective testing results. We conclude the paper with a section summarizing the accomplishments of the project and also identify potential next steps and recommendations based on the technical results and engineering experience gained throughout the execution of the VSC-A Project.

INTRODUCTION

Vehicle-to-Vehicle (V2V) safety communications can play a major role in addressing vehicle crashes where multiple vehicles are involved. According to [1], this technology can reduce, mitigate, or prevent 82 percent of crashes by unimpaired drivers. The communications technology for V2V is 5.9 GHz Dedicated Short Range Communications (DSRC). This wireless communications technology has a very low latency and is considered to be the technology of choice for the types of crash avoidance applications that were prototyped in the Vehicle Safety Communications-Applications (VSC-A) Project [2]. The major objectives of the VSC-A development activities were the:

- Selection of high-value safety applications
- Development of a test bed that allowed interoperability between different car manufacturers
- Development and standardization of a message set for vehicle safety communications
- Development of an accurate relative positioning system
- Prototyping of safety applications
- Objective testing of the safety applications

A primary goal of the VSC-A Project was to determine whether systems that utilized DSRC-based V2V communications and positioning can help overcome limitations of autonomous systems and enhance the overall performance of safety systems. One potential advantage of V2V safety communications is that it may provide significant, additional information about the driving situation and expand the awareness horizon of the vehicle well beyond the capabilities of vehicle-autonomous sensors. Another advantage of V2V systems is that it may be possible to integrate such systems on vehicles in which the system was not original equipment, including retrofit of existing vehicles.

In order to gauge the feasibility of such systems, a reference system and applications to address crash imminent scenarios were implemented. This reference system ("test bed") combined communications, accurate relative positioning and security and was integrated with the vehicles from the five Original Equipment Manufacturers (OEMs) that participated in the VSC-A Project. A fundamental aspect of the project was the establishment of interoperability between different OEMs. This interoperability requirement led to the development of the V2V message set, which was standardized in SAE J2735 as the Basic Safety Message (BSM) [3]. The development of the test bed and the applications followed a systems engineering process and the resulting minimum performance requirements formed the basis for the development and the testing of the applications. To test the performance of the test bed and the applications, objective test procedures were developed together with the United States Department of Transportation (USDOT) and the testing was performed at the Transportation Research Center (TRC) in East Liberty, Ohio with the aid of the National Highway Traffic Safety Administration's (NHTSA) Vehicle Research and Test Center (VRTC).

CRASH SCENARIOS AND APPLICATION SELECTION

To provide a foundation for the VSC-A Project, the USDOT evaluated pre-crash scenarios based on the 2004 General Estimated Systems (GES) crash database. This list served as the basis for the selection of the safety applications to be prototyped under the VSC-A Project. Each crash scenario was assigned a composite crash ranking determined by taking the average of the crash rankings by frequency, cost, and functional years lost for each scenario. The crash scenarios were then sorted based on the composite ranking and were analyzed to evaluate whether autonomous safety systems and/ or vehicle safety communications would offer the best opportunity to adequately address the scenarios.

From this ranked list of crash scenarios (based on crash frequency, crash cost and functional years lost) the top seven (7) crash scenarios to be addressed by the VSC-A Project were selected. The selected crash-imminent scenarios were analyzed and potential, DSRC-based, safety application concepts capable of addressing them were developed. The crash imminent scenarios and the applications selected to be part of the VSC-A safety system is shown in Table 1. The VSC-A team together with the USDOT analyzed the scenarios in Table 1 and developed concepts for safety applications that could potentially address them through vehicle safety communications. This analysis resulted in the identification of the following safety applications as part of the VSC-A system:

Emergency Electronic Brake Lights (EEBL), defined as follows

The EEBL application enables a host vehicle (HV) to broadcast a self-generated emergency brake event to surrounding remote vehicles (RVs). Upon receiving the event information, the RV determines the relevance of the event and issues a warning to the driver, if appropriate. This application is particularly useful if the drivers' line of sight is obstructed by other vehicles or bad weather conditions (e.g., fog, heavy rain)

Forward Collision Warning (FCW), defined as follows

The FCW application is intended to warn the driver of the HV of an impending rear-end collision with an RV ahead in traffic in the same lane and direction of travel. FCW is intended to help drivers in avoiding or mitigating rear-end vehicle collisions in the forward path of travel.

Blind Spot Warning+Lane Change Warning (BSW +LCW), defined as follows

The BSW+LCW application is intended to warn the driver during a lane change attempt if the blind-spot zone into which the HV intends to switch is, or will soon be, occupied by another vehicle traveling in the same direction. Moreover, the application provides advisory information that is intended to inform the driver of the HV that a vehicle in an adjacent lane is positioned in a blind-spot zone of the HV when a lane change is not being attempted.

Do Not Pass Warning (DNPW), defined as follows

The DNPW application is intended to warn the driver of the HV during a passing maneuver attempt when a slower moving vehicle, ahead and in the same lane, cannot be safely passed using a passing zone which is occupied by vehicles in the opposite direction of travel. In addition, the application provides advisory information that is intended to inform the driver of the HV that the passing zone is occupied when a vehicle is ahead and in the same lane and a passing maneuver is not being attempted.

Intersection Movement Assist (IMA), defined as follows

The IMA application is intended to warn the driver of a HV when it is not safe to enter an intersection due to high collision probability with other RVs. Initially, IMA is intended to help drivers avoid or mitigate vehicle collisions at stop sign-controlled and uncontrolled intersections.

Control Loss Warning (CLW), defined as follows

The CLW application enables a HV to broadcast a self-generated, control-loss event to surrounding RVs. Upon receiving such event notification, the RV determines the relevance of the event and provides a warning to the driver, if appropriate.

Table 1 illustrates the mapping between the crash imminent scenarios and the safety applications defined above.

DEVELOPMENT OF THE TEST BED

Each OEM in the VSC-A Project developed a vehicle test bed to serve as a prototype platform for the V2V communications system. The OEMs jointly developed system specifications and performance requirements that served as the basis for the system and application developments. The test bed was based on a common prototype platform referred to as the On-Board Equipment (OBE). The selected OBE allowed development flexibility and was representative of current (or future) automotive grade processing power. The OBE contained a DSRC radio, a processor and various interfaces (e.g., for vehicle data, Global Positioning System (GPS) data, etc.). The test bed was an effective tool for validating safety application concepts, system test procedures and for answering critical research questions regarding V2V communications. Those issues included relative lane-level positioning, time synchronization, communications scalability and practical security and anonymity.

SOFTWARE ARCHITECTURE

In order to support the functionality of the safety applications described earlier and their development, the activities initially focused on the development of a system architecture based on various modules that could be upgraded independently from each other, if necessary. This approach allowed for fast and efficient prototyping throughout the development phase of the project. This architecture was used during the test bed design stage for the definition of the Hardware (HW) and Software (SW) architectures and required interfaces. The various modules forming the system test bed were categorized into the following major groups: Interface, Positioning & Security, Core, Safety Applications, Threat Process and Reporting, and Data Analysis. The system block diagram (Figure 1) shows the breakdown of the individual modules that make up each of the major module groupings. This provided a good framework for a comprehensive V2V safety system.

The focus of the system design activities was the core modules (Target Classification, Host Vehicle Path Prediction and Path History) and the positioning, security and safety application modules. The system design was based on the preliminary requirement specifications developed for each of the modules. Testing of the system resulted in updates to the modules throughout the project, culminating in the final test bed implementation. In the next section the software modules are described briefly.

SOFTWARE MODULES

The VSC-A software modules are composed of support and application functions. The support functions provide the interface to any external equipment and they calculate the necessary parameters to support the application modules and the Engineering Driver-Vehicle Interface (DVI). The primary software modules are:

- Threat Arbitration (TA)
- Driver-Vehicle Interface Notifier (DVIN)
- Target Classification (TC)
- Host Vehicle Path Prediction (HVPP)
- Path History (PH)
- Data Logger (DL)
- Engineering Graphical User Interface (EGUI)
- Sensor Data Handler (SDH)
- Wireless Message Handler (WMH)

The application modules evaluate potential safety threats based on the data and inputs from the support modules. The application modules contain the warning algorithms for the safety applications shown in Table 1. The SDH and WMH are basic, functional blocks necessary for parsing inputs from and submitting data to the software services of the system platform and those in use by the other support and application elements. The SDH interfaces to the vehicle Controller Area Network (CAN) gateway device to transmit and receive CAN messages and detect communication errors. It also connects to the GPS receiver to obtain National Marine Electronics Association (NMEA) data including Universal Coordinated Time (UTC) time, position, speed and heading, as well as raw GPS data. The SDH also interfaces to the external computing platform that executes the Real Time Kinematic (RTK) software to obtain accurate relative positioning of the neighboring vehicles. The WMH interfaces to the DSRC radio and to the Security Module (SM) software. It transmits and receives WAVE Safety Messages (WSM) using the SM to generate and verify message signatures. The TC categorizes the surrounding vehicles according to their position and heading relative to the HV, using the HVPP and the PH of the HV and RV. The TA arbitrates between multiple threats and chooses the one with the highest crash energy as the one to display to the driver and sends the respective request to the DVIN, which activates the corresponding alert in the EGUI.

The VSC-A team decided to use the shared memory interface concept. This allows for data in memory to be accessed by multiple modules for inter-process communication. This is advantageous, because there are many cases of one module supplying data to other functional blocks. For example, consecutive host and remote GPS time and position data points may be used by HVPP, PH, TC and the warning algorithms at the same time. The shared memory scheme used in the architecture fulfills the requirements for support of the VSC-A functionality while allowing for extensibility of the architecture.

ENGINEERING GUI

The EGUI is an "engineering-type" graphical user interface with the purpose to provide a simple engineering tool that could be used to understand, evaluate, and configure the VSC-A platform. It allows representation of visual and auditory vehicle driver warnings as a result of the application module processes. The touch-screen interface also allows the user to view and control parameters necessary for the operation of the VSC-A safety applications. Figure 2 shows examples of the graphical interface as depicted on a Video Graphics Array (VGA) touch screen.

This allowed the EGUI to display the warning states of a particular threat (e.g., DNPW in Figure 2). Only one of the warning screens is visible at any particular time. In order to ensure that the most important warning was shown on the DVI screen, the TA uses the threat level, relative speed, and location of the threat from each of the application modules to assess the severity and determine the highest priority request to be used by DVIN.

IN-VEHICLE HARDWARE INTEGRATION

The in-vehicle HW integration involved the selection, purchasing, installation and integration of all the HW and SW required for completion of the test bed. Table 2 identifies the model and manufacturer of the equipment installed on the VSC-A test bed vehicles.

MESSAGING STANDARDS

A major goal of the VSC-A Project was to define a single Over-the-Air (OTA) message whose contents could support all of the VSC-A safety applications as well as other safety applications that are likely to be developed in the future. That goal was achieved with the standardization of the SAE J2735 BSM [3]. An internal version of the OTA message was defined and implemented in the test bed with the objective testing verifying that this message supports all of the VSC-A applications. The BSM consists of Parts I and II. A proposal was prepared and presented for SAE to redefine both Parts I and II of the BSM. Part I consists of vehicle state data that is so critical for safety applications that it must be included in every BSM. Part II consists of data that is either required by applications at regular intervals (potentially at a reduced frequency), required to notify applications of a given event or optional for applications. Figure 3 shows the components and format of the BSM in SAE J2735.

The SAE J2735 conformant BSM uses the Distinguished Encoding Rules (DER) to encode the message for OTA transmission. In addition to the effort to develop and standardize the BSM, the VSC-A team also initiated a new SAE DSRC standards project (SAE J2945) for BSM minimum performance requirements. This standard will augment SAE J2735 to define rules necessary for effective V2V safety communications interoperability (e.g., minimum message rate, minimum data accuracy, etc.).

OBJECTIVE TESTING

OVERVIEW

The objective testing activity included the development of the Objective Test Procedures (OTPs) and test plan, conducting the objective tests, and analyzing the test results. The purpose of the objective testing was to ascertain that:

- The performance of the VSC-A system test bed was sufficient to enable the safety applications in the project
- The safety applications satisfied the minimum performance requirements developed in the system design activity of the project

The OTPs were developed for each application and were designed to include the most common scenarios that the application would encounter. The procedures included the following:

- True positive tests, where the objective is to get a warning
- False positive tests, where the objective is to suppress a warning because it is not needed

The outcomes of the objective tests were used by the Volpe National Transportation Systems Center (Volpe) to estimate the safety benefits opportunity for V2V communications based safety applications. In total, 33 test procedures were developed, 22 true positive tests and 11 false positive tests. For the benefits estimate, only the true positive tests which all had successful/unsuccessful criteria associated with them were evaluated. The OTPs were discussed with NHTSA and Volpe and agreed upon by all the participants. Following the OTP development, the test plan was written. It included the number of runs for each test, test speeds, validation criteria for each test (allowable speed ranges, etc.) and detailed setup procedures to make the OTPs as repeatable as possible. The test plan was also agreed upon by Volpe and NHTSA prior to the start of testing. The objective testing took place from June 1, 2009 to June 3 2009 at TRC in East Liberty, Ohio.

The data that was collected during the testing was recorded in a data logging and visualization tool called CANape [4]. CANape is a SW tool developed by Vector CANtech, Inc. and is used for the development, calibration and diagnostics of Electronic Control Units (ECUs) as well as data acquisition and analysis. The CANape software was customized by Vector for the VSC-A Project. Figure 4 shows an example of the primary screen that was used for the objective testing. The screen is divided into four quadrants as follows:

- Quadrant 1 contains a birds-eye view, which is a graphical representation of the location of the HV, centered at (0,0) and the RVs that the HV is in communication with
- Quadrant 2 contains the camera data which will consist of a single image, as shown below, or up to four images multiplexed together
- Quadrant 3 contains the HV's sensor data and GPS data
- Quadrant 4 contains the RV track data as determined by the TC core module

OBJECTIVE TEST RESULTS

The complete list of tests, the speed for the runs, the number of runs for each test and the test outcome is shown in Table 3. As can be seen from Table 3, all the applications passed the objective tests.

SUMMARY/CONCLUSIONS

The major accomplishments of the project are:

- Defined a set of high-priority, potential crash scenarios that could be addressed by V2V communication
- Selected and developed a set of V2V applications to address the above set of potential crash scenarios
- Defined efficient system architecture for V2V safety system where all VSC-A safety applications are enabled at the same time
- Successfully implemented a test bed with all the safety applications on a platform running an automotive grade processor (400 MHz)
- Successfully incorporated and evaluated in the test bed two relative positioning approaches (RTK and Single Point (SP))
- Successfully incorporated in the test bed the necessary OTA communication protocol (SAE J2735) and security protocol (IEEE 1609.2 Elliptic Curve Digital Signature Algorithm (ECDSA) [5] with Verify-on-Demand (VoD) [6])
- Defined OTPs for all the VSC-A safety applications, including true positive and false positive tests
- Successfully executed and passed all objective tests for all the VSC-A safety applications

• Refined, with field data, the required OTA message set for V2V safety (BSM within SAE J2735) which led to the recently published version of the standard [3]

• Conducted a study to quantify availability and accuracy of GPS-based relative positioning by using RTK and SP methods for V2V

• Confirmed that IEEE 1609.2 ECDSA with VoD functioned properly under all test conditions for the VSC-A safety applications

• Performed and analyzed initial scalability with up to 60 radios [8] to characterize channel behavior under IEEE 1609.4 [7] and under dedicated full time use of channel 172

Another outcome of the technical work was the identification of technical questions and topics that still need to be answered for any successful deployment:

• How does the system perform with large numbers of communicating nodes?

• How can security certificates be managed and privacy preserved?

• Are the standards sufficient for interoperability?

• What are requirements for data reliability and integrity?

• What are technical solutions for acceleration of market penetration?

• How to enhance the safety applications and system design?

• How to enhance relative vehicle positioning?

Those questions and topics are being addressed under the current NHTSA V2V safety roadmap [1] which outlines the next set of activities needed to support a NHTSA decision regarding V2V safety in 2013.

REFERENCES

1. DOT-Sponsored Research Activities: V2V Communications for Safety, http://www.intellidriveusa.org/research/v2v.php.

2. Vehicle Safety Communications - Applications (VSC-A) - First Annual Report, http://www.intellidriveusa.org/documents/2009/05/09042008-vsc-a-report.pdf

3. SAE International Surface Vehicle Standard, "Dedicated Short Range Communications (DSRC) Message Set Dictionary," SAE Standard J2735, Rev. Nov. 2009.

4. CANape, A Versatile Tool for Measurement, Calibration and Diagnostics of ECUs, Vector, http://www.vector.com/vi_canape_en.html?quickfinder=1.

5. IEEE Trial-use Standard 1609.2TM-2006, WAVE - Security Services for Applications and Management Messages, 2006.

6. Krishnan, H., Technical Disclosure, "Verify-on-Demand" - A Practical and Scalable Approach for Broadcast Authentication in Vehicle Safety Communication, IP.com number: IPCOM000175512D, IP.com Electronic Publication: October 10, 2008.

7. IEEE P1609.4TMD6.0, Draft Standard for Wireless Access in Vehicular Environments - Multi-channel Operation, IEEE Vehicular Technology Society, March 2010.

8. Ahmed-Zaid, F., Krishnan, H., Maile, M., Caminiti, L. et al., "Vehicle Safety Communications - Applications: Multiple On-Board Equipment (OBE) Testing," *SAE Int. J. Passeng. Cars – Mech. Syst.* 4(1):547-561, 2011, doi: 10.4271/2011-01-0586.

CONTACT INFORMATION

Farid Ahmed-Zaid
Ford Motor Company
fahmedza@ford.com

ACKNOWLEDGMENTS

The CAMP VSC2 Participants would like to acknowledge the following USDOT personnel for their invaluable project support; Art Carter, Ray Resendes, and Mike Schagrin. The Participants would also like to thank VRTC personnel, especially Garrick Forkenbrock, for their outstanding support during the execution of the objective tests. Finally the Participants would like to express their appreciation to the following Volpe personnel; Wassim Najm, Bruce Wilson, and Jonathan Koopman for their support with the development and execution of the objective tests.

DEFINITIONS/ABBREVIATIONS

BSM
Basic Safety Message

BSW/LCW
Blind Spot Warning, Lane Change Warning

CAMP
Crash Avoidance Metrics Partnership

CAN
Controller Area Network

CLW
Control Loss Warning

DER
Distinguished Encoding Rules

DL
Data Logger

DNPW
Do Not Pass Warning

DSRC
Dedicated Short Range Communications

DVI
Driver-Vehicle Interface

DVIN
Driver-Vehicle Interface Notifier

ECDSA
Elliptic Curve Digital Signature Algorithm

ECU
Electronic Control Unit

EEBL
Emergency Electronic Brake Lights

EGUI
Engineering Graphical User Interface

FCW
Forward Collision Warning

GES
General Estimated Systems

GPS
Global Positioning System

HV
Host Vehicle

HVPP
Host Vehicle Path Prediction

HW
Hardware

IMA
Intersection Movement Assist

ITS
Intelligent Transport Systems

NHTSA
National Highway Traffic Safety Administration

NMEA
National Maritime Electronics Association

OBE
On-Board Equipment

OEM
Original Equipment Manufacturer

OTA
Over-the-Air

OTP
Objective Test Procedure

PH
Path History

RTK
Real-Time Kinematic

RV
Remote Vehicle

SDH
Sensor Data Handler

SM
Security Module

SP
Single Point (positioning)

SW
Software

TA
Threat Arbitration

TC
Target Classification

TRC
Transportation Research Center

USDOT
United States Department of Transportation

UTC
Universal Coordinated Time

V2V
Vehicle-to-Vehicle

VGA
Video Graphics Array

VoD
Verify-on-Demand

VRTC
Vehicle Research and Test Center

VSC2
Vehicle Safety Communications 2 (Consortium)

VSC-A
Vehicle Safety Communications - Applications

WMH
Wireless Message Handler

WSM
Wave Short Message

DISCLAIMER

This material is based upon work supported by the National Highway Traffic Safety Administration under Cooperative Agreement No. DTNH22-05-H-01277. Any opinions, findings, and conclusions or recommendations expressed in this publication are those of the Author(s) and do not necessarily reflect the view of the National Highway Traffic Safety Administration.

APPENDIX A

OBJECTIVE TEST PROCEDURE EXAMPLE AND TEST RESULTS

In this appendix we provide an example of the test plan and OTP together with the results of the testing. The chosen example is the FCW, Test 1.

FCW OBJECTIVE TEST PROCEDURES

FCW is a V2V, communication-based, safety feature that issues a warning to the driver of the HV in case of an impending rear-end collision with a vehicle ahead in traffic in the same lane and direction of travel. FCW is designed to help drivers in avoiding or mitigating rear-end vehicle collisions in the forward path of travel.

FCW-T1: HV Travel at a Constant Speed to a Stopped RV

Background

This test begins with the HV traveling on a straight, flat road at 50 mph. Ahead of the HV, in the same lane, is a single RV stopped in the lane of travel. The test determines whether the countermeasure's required collision alert occurs at the expected range. This test especially explores the ability of the countermeasure to accurately identify stationary in-path targets on a flat, straight road.

Test Setup

Figure 5 shows the vehicle positions and test setup for Test 1.

Cones with flags are placed so the driver of the HV is aware of the vehicle's location in reference to the required maneuvers. These flags are located by their distance from the starting point for the HV. It is assumed that flags will be placed using an accurate GPS handheld receiver. Alternate methods of flag location can be used. Flag locations are:

- A red flag is placed at the starting point where the HV begins its maneuver (cone not shown)
- A yellow flag is placed at the point where the HV reaches the target speed (cone HV-A), at least 650 meters from the red flag
- A white flag is placed at the earliest valid (from the driver's perspective) WARN point (cone HV-B)

A checkered flag is placed where the HV will make an evasive maneuver by changing lanes if the WARN has failed to occur (cone HV-C) which is positioned at 90 percent of the allowable alert range. At the test speed of 50 mph, this is 9 meters from HV-B cone

A green flag is placed at the stopping position for the RV (cone RV-A), at least 800 meters from the red flag

Driving Instructions

- The RV begins at the starting point and stops with its front bumper at the green flag
- The HV starts accelerating at least 800 meters behind the RV in the same lane to reach a speed of 50 mph
- The HV Cruise Control is set at the required speed of 50 mph
- The HV Cruise Control shall be engaged at least 150 meters behind the RV
- The warning will be given at around the nominal warn range (cone HV-B) after which the HV will change lane

[Note: If the warning is not given when the HV reaches the checkered flag (cone HV-C), the HV shall make an evasive maneuver by changing lanes and come to a safe stop in the adjacent lane.]

Successful Criteria

- The collision alert shall occur within the ranges specified in Table 4 in order to pass the run
- If at least six runs out of eight runs pass, then the test is successful

Unsuccessful Criteria

- A run is unsuccessful if any of the conditions below occur:
    - Collision alert occurrence outside the range calculated in Table 4 using run-specific variables
    - The warning is missed such that the HV passes cone HV-C and no alert is triggered
    - If at least three runs out of eight runs fail, the test is unsuccessful

Table 4. Alert Range for FCW Test 1

	Collision Alert Test
Maximum Range	93.7
Nominal Range	85.2
Minimum Range	76.7

Evaluation Criteria

Number of Valid Test Runs	HV Speed (mph)	RV Speed (mph)	Number of Successful Test Runs
8	50	0	≥ 6

FCW OBJECTEIVE TEST 1 RESULTS

For the FCW application to pass, the warning had to come between the maximum and minimum alert range that was calculated for each run. As can be seen from the test results table (Table 5), the application was successful in all the runs for the test.

APPENDIX B

TABLES AND FIGURES

Table 1. Mapping of VSC-A Program Applications to Crash Imminent Scenarios

	V2V Safety Applications/ Crash Scenarios	EEBL	FCW	BSW	LCW	DNPW	IMA	CLW
1	Lead Vehicle Stopped		✓					
2	Control Loss without Prior Vehicle Action							✓
3	Vehicle(s) Turning at Non-Signalized Junctions						✓	
4	Straight Crossing Paths at Non-Signalized Junctions						✓	
5	Lead Vehicle Decelerating	✓	✓					
6	Vehicle(s) Not Making a Maneuver – Opposite Direction					✓		
7	Vehicle(s) Changing Lanes – Same Direction			✓	✓			

Table 2. VSCA Test Bed Hardware List

Item Description	Manufacturer	Model
GPS Receiver	NovAtel®	OEMV® Flexpak V1-RT20A
GPS Antenna	NovAtel®	GPS-701-GG
LCD VGA Monitor	Xenarc	700TSV-B
USB CCD Monochrome Camera	The Imaging Source	DMK 21BU04
Car PC	Logic Supply	Voom PC-2
Inertial Measurement Unit	Silicon Sensing	DMU
OBE Vehicle CAN interface	Smart Engineering Tools	Netway 6
DSRC Antenna	Nippon Antenna	DEN-HA001-001
Ethernet Switch	Netgear	GS105

Table 3. Objective Test Scenarios and Results

Test Scenario	Description	Speeds	Number of Runs	Type of Test	Result
EEBL-T1	HV at constant speed with decelerating RV in same lane	50	8	True Positive	Successful
EEBL-T2	HV at constant speed with decelerating RV in left lane on curve	50	8	True Positive	Successful
EEBL-T3	HV at constant speed with decelerating RV in same lane and obstructing vehicle in between	50	8	True Positive	Successful
EEBL-T4	HV at constant speed with mild-decelerating RV in same lane	50	2	False Positive	N/A

Test Scenario	Description	Speeds	Number of Runs	Type of Test	Result
EEBL-T5	HV at constant speed with decelerating RV in 2nd right lane	50	2	False Positive	N/A
FCW-T1	HV travel at a constant speed\RV stopped	50	10	True Positive	Successful
FCW-T2	HV travel behind RV1\RV1 travel behind RV2\RV2 stopped	50	10	True Positive	Successful
FCW-T3	HV drive on a curve\RV stopped at the curve	50	8	True Positive	Successful
FCW-T4	HV tailgate RV	50	2	False Positive	N/A
FCW-T5	HV follows RV\RV brakes hard	40	10	True Positive	Successful
FCW-T6	HV driving into a curved right lane\RV stopped in the left curved lane	50	2	False Positive	N/A
FCW-T7	HV travels behind a slower RV	50	10	True Positive	Successful
FCW-T8	HV changes lanes behind a stopped RV	50	8	True Positive	Successful
FCW-T9	HV approaches two RVs in left and right adjacent lanes and passes between them	50	2	False Positive	N/A
BSW/LCW-T1	LCW Warning, Left	50	8	True Positive	Successful
BSW/LCW-T2	LCW Warning, Right	50	8	True Positive	Successful
BSW/LCW-T3	LCW Warning, Right with Left BSW Advisory	50	9	True Positive	Successful
BSW/LCW-T4	BSW Advisory Alert, Left	50	8	True Positive	Successful
BSW/LCW-T5	BSW Advisory Alert, Right	50	8	True Positive	Successful
BSW/LCW-T6	No Warning or Advisory for RV behind	50	2	False Positive	N/A

Test Scenario	Description	Speeds	Number of Runs	Type of Test	Result
BSW/LCW-T7	No Warning or Advisory for RV far Right	50	2	False Positive	N/A
BSW/LCW T8	LCW Warning in Curve, Right	35	8	True Positive	Successful
DNPW-T1	Attempt to pass with oncoming RV in adjacent lane	25/35	10	True Positive	Successful
DNPW-T2	Attempt to pass with stopped RV in adjacent lane	30/40	10	True Positive	Successful
DNPW-T3	Attempt to pass with oncoming RV not in adjacent lane	45	2	False Positive	N/A
IMA-T1	Variable speed approaches with stopped HV/moving RV/open intersection	20/30/40/50	12	True Positive	Successful
IMA-T2	Stopped HV/moving RV/open intersection	35/50	4	False Positive	N/A
IMA-T3	Variable speed approaches with moving HV/moving RV/open intersection	15/25/35/45	16	True Positive	Successful
IMA-T4	Moving HV/moving RV/open intersection	25	4	False Positive	N/A
IMA-T5	Stopped HV/moving RV/open intersection/parked vehicle	20/30/40/50	12	True Positive	Successful
CLW-T1	HV at constant speed with CLW RV in same lane ahead in same travel direction	40	8	True Positive	Successful
CLW-T2	HV at constant speed with CLW RV in 2nd right lane	30	2	False Positive	N/A
CLW-T3	HV at constant speed with CLW RV in adjacent lane ahead in opposite travel direction	30	12	True Positive	Successful

Table 5. FCW Test 1 Results

Run	Actual Values at Alert Onset: HV Speed	HV Accel	RV Speed	RV Accel	Calculated Run-Specific Alert Ranges: Maximum (meters)	Nominal (meters)	Minimum (meters)	Actual Alert Range (meters)	Pass/Fail	Headway
FCW T1	50 mph 22.35 m/s	0 g	0 mph	0 g	93.7	85.2	76.7			6.71
1	49.70 mph 22.21 m/s	0 g	0 mph	0 g	92.8	84.4	75.9	84	Pass	6.80
2	50.0 mph 22.35 m/s	0 g	0 mph	0 g	93.7	85.2	77.7	86	Pass	6.83
3	49.68 mph 22.21 m/s	0 g	0 mph	0 g	91.7	83.4	75.0	85	Pass	6.76
4	50.31 mph 22.49 m/s	0 g	0 mph	0 g	94.1	85.5	77.0	85	Pass	6.80
5	49.69 mph 22.21 m/s	0 g	0 mph	0 g	92.8	84.4	75.9	85	Pass	6.80
6	49.50 mph 22.13 m/s	0 g	0 mph	0 g	93.1	84.6	76.2	80	Pass (A # of communication outages from the RV)	6.82
7	49.84 mph 22.28	0 g	0 mph	0 g	93.9	85.4	76.8	83	Pass	6.73

	Actual Values at Alert Onset				Calculated Run-Specific Alert Ranges					
Run	HV Speed	HV Accel	RV Speed	RV Accel	Maximum (meters)	Nominal (meters)	Minimum (meters)	Actual Alert Range (meters)	Pass/Fail	Headway
	m/s									
8	49.75 mph 22.24 m/s	0 g	0 mph	0 g	92.7	84.3	75.9	82	Pass (A # of communication outages from the RV)	6.79
9	50.35 mph 22.51 m/s	0 g	0 mph	0 g	93.2	84.7	76.3	85	Pass	6.66
10	49.71 mph 22.22 m/s	0 g	0 mph	0 g	91.7	83.3	75.0	84	Pass	6.75

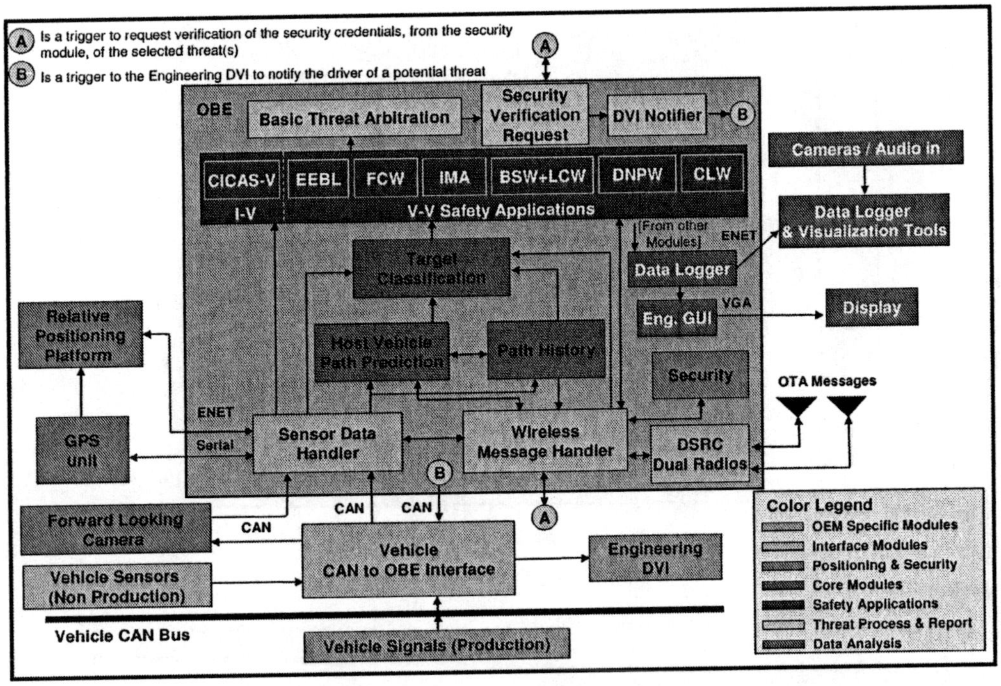

Figure 1. VSC-A System Block Diagram

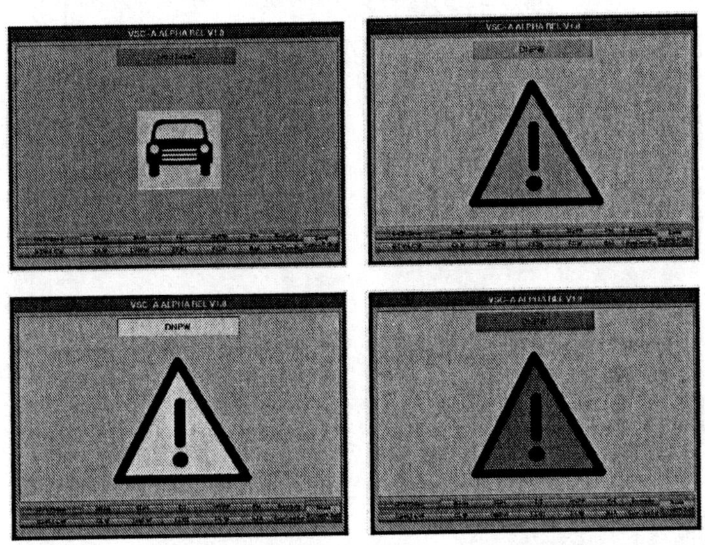

Figure 2. DVIN Stages (left to right, top to bottom) No Threat, Threat Detected, Inform Driver, Warn Driver

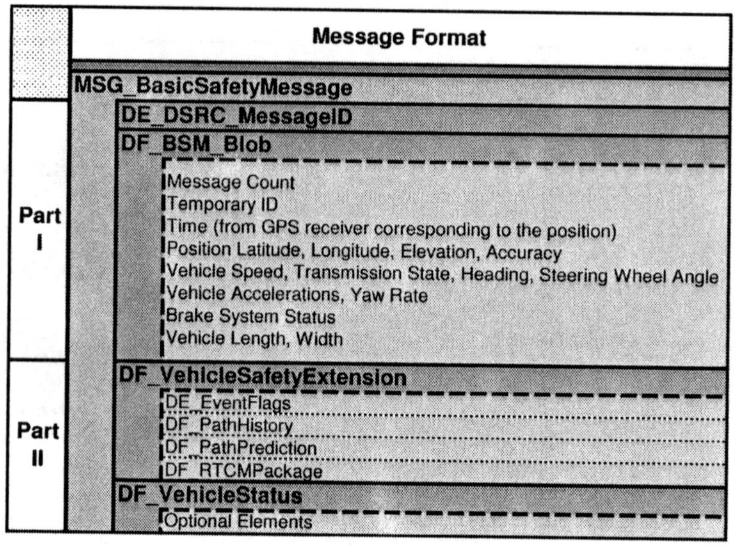

Figure 3. SAE J2735 Rev 35 Basic Safety Message Format

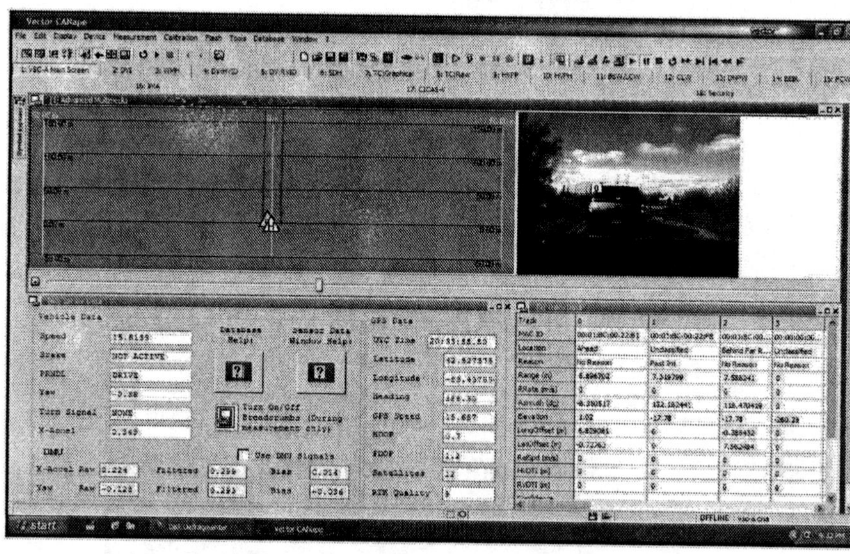

Figure 4. Example Layout Screen for OTP Testing

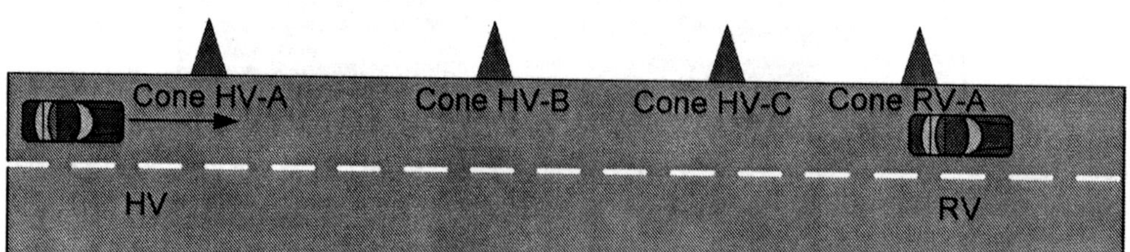

Figure 5. FCW Test 1 Test Setup - RV in Same Lane

A Sampling of Autonomous Vehicle Research Projects

SAE International

Distributed System Architecture of Autonomous Vehicles and Real-Time Path Planning based on the Curvilinear Coordinate System

2012-01-0740
Published
04/16/2012

Keonyup Chu, Junsoo Kim and Myoungho Sunwoo
Hanyang Univ

doi:10.4271/2012-01-0740

ABSTRACT

The development of autonomous vehicle requires the state-of-the-art technologies in perception, planning, control, and system integration. This paper presents an overview of the system architecture and software architecture of autonomous vehicles for system integration. Network based system architecture in this paper provides a distributed computing system for autonomous driving. Further, a real-time path planning and a target speed generation are described based on the curvilinear coordinate system. The design of a path in the curvilinear coordinate system stretches the design space as like the Cartesian coordinate system to simplify the generation of the path. In determination of target speed, curvatures and risk of a generated path were utilized for safe autonomous driving. The proposed system architecture and planning algorithm were successfully integrated into the autonomous vehicle A1, which was developed by the Automotive Control and Electronics Laboratory (ACE Lab) and Machine Monitoring and Control Laboratory (MMC Lab) from Hanyang University. Seoul, Korea.

INTRODUCTION

For the sake of development of the human-friendly vehicle, several technological issues have been addressed focusing on the improvement of the driving safely and the convenience as a living space. In improving these issues for safely and convenience of the vehicle, intelligent vehicle technologies have played an important role by realizing the functionality and commercialization for human-friendly vehicle. Intelligent vehicle technologies still have unsolved matter dramatic potential to improve driving safely and the convenience of the vehicle. Eventual evolution of the intelligent vehicle leads to the autonomous vehicle [1, 2]. Autonomous driving capability enables the driver to do other tasks or to take a rest while driving and it maximizes the convenience of the life in the vehicle. Application of the autonomous driving system will reduce labor costs for public transportation system like bus or for commercial vehicles like taxi or truck. Moreover, the autonomous vehicle can achieve ultimate safely of the driving by removing human errors and perceptual limitations from the vehicle operation control loop. Essentially, driving of a vehicle consists of three basic functions: perception, planning, and control. Perception function is an information extraction process from an environment for the utilization to the driving. And the planning is a decision process for behavior of the vehicle. Control is responsible for actions of the vehicle such as turning, acceleration and braking [1, 2]. Among the three basic functions of the driving, the planning allows the autonomous vehicle to decide the behavior of vehicle by itself. The path planning is responsible for the obstacle avoidance of an autonomous vehicle following predefined global route [3, 4, 5, 6]. Therefore, the primary research objective of this paper is the design of the path planner for the autonomous vehicle. Further, the integration of the entire sub-systems is also important issue for the robustness and reliability of the autonomous vehicle. In order to address this issue, the system architecture and software architecture were carefully designed and implemented. The proposed autonomous driving system and planning algorithms were applied to the autonomous vehicle A1 [7], which won the 2010 Autonomous Vehicle Competition (AVC) organized by the Hyundai Motor Group in Korea [8]. The autonomous vehicle A1 developed by the Automotive Control and Electronics Laboratory (ACE Lab) and Machine Monitoring and Control Laboratory (MMC Lab) from Hanyang University, Seoul, Korea is shown in Figure 1.

Figure 1. A1: the autonomous vehicle of Hanyang University in the AVC

SYSTEM ARCHITECTURE

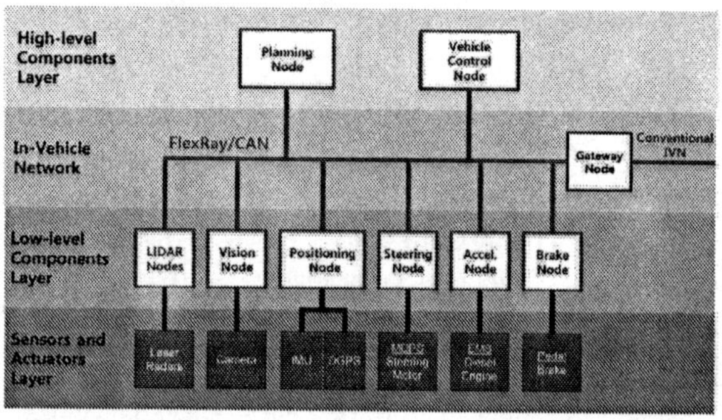

Figure 2. System Architecture of A1

Autonomous vehicle A1 is developed based on a Hyundai Tucson ix which is powered by the 2.0 liter diesel engine. The Tucson ix is a sports utility vehicle that has an electric power steering (EPS) system and an electronic stability control (ESC). For obstacle detection, two laser scanners looking down the road ahead were installed on the roof of the autonomous vehicle Al, and three laser scanners scanning parallel to the vehicle were mounted on the bumper. For the measurement of the global vehicle position, two Differential Global Positioning Systems (DGPSs) were installed on the roof of the A1. An inertial measurement unit (IMU) mounted on the interior of the vehicle, steering angle sensor, four wheel speed sensors, and two DGPSs are all used for the configuration of the positioning system [9]. In order to detect the lane and the crosswalk on the on-road, the single camera is located on the front windshield inside the vehicle. The autonomous driving system of A1 has a layered architecture based on the In-Vehicle Network (IVN) as shown in Figure 2: high-level component layer, In-Vehicle Network, low-level components layer, and sensor and actuator layer. This network based layered architecture provides a distributed computing system architecture of autonomous driving system.

HIGH-LEVEL COMPONENT LAYER

The role of high-level components layer is the integration of entire information from perception nodes and to make decisions. This layer consist of two nodes: planning node and vehicle control node. The planning node receives the vehicle states and environment information from the low-level components layer through the IVN, and it generates a drivable path and target vehicle speed. In the vehicle control node, the practical control commands such as desired steering angle, acceleration position, and brake position are computed using the path given by planning node. These decisions are transmitted to the steering node, the acceleration node, and the brake node in the low-level components layer through the IVN.

LOW-LEVEL COMPONENT LAYER

Low-level components are composed of several nodes to process raw data from the sensors and also to control the actuators. Low-level components layer has three kinds of perception nodes: LIDAR node, vision node, and positioning node. For the control of vehicle motion, the steering node, the acceleration node, and the brake node are connected to individual actuator.

SENSOR AND ACTUATOR COMPONENTS LAYER

The sensor and actuator components layer are composed of laser scanners, camera, IMU, DGPS, EPS, diesel engine, and brake.

IN-VEHICLE NETWORK

In a distributed real-time system, the system integration is accomplished by communications between the different nodes. For this reason, the IVN has a significant role in the integration of the distributed autonomous driving system. The IVN consists of Controller Area Network (CAN), FlexRay, Gateway, and conventional IVN of Tucson ix. In our application, CAN is used as the primary network protocol for the communication between multiple nodes of the autonomous driving systems and FlexRay is used for the purpose of the synchronization between LIDAR nodes [10, 11]. The purpose of gateway in our application is to transmit the vehicle sensor data such as steering angle and wheel speed from the conventional IVN of Tucson ix to several nodes in the autonomous driving system.

SOFTWARE ARCHITECTURE

The software of the A1 consists of five groups which correspond to the following functions: sensor interface, perception, navigation, actuator interface, and user interface. The sensor interface and actuator interface are hardware dependent software modules that manage interaction with the sensors and actuators. In the other hand, the perception and navigation consist of hardware independent software modules that play significant role in realizing the autonomous driving by processing the information. The processing of information is executed by each software module that is represented as a block in Figure 3. The information in the entire software architecture flows from the sensors to the actuators in single direction. The opposite direction of data flow is not allowed except for the user interface to simplify software design and to prevent the deadlock between each module. This modular software architecture enables each software module of the subsystem to be individually developed and tested. For this reason, the management and debug of entire software was efficiently performed.

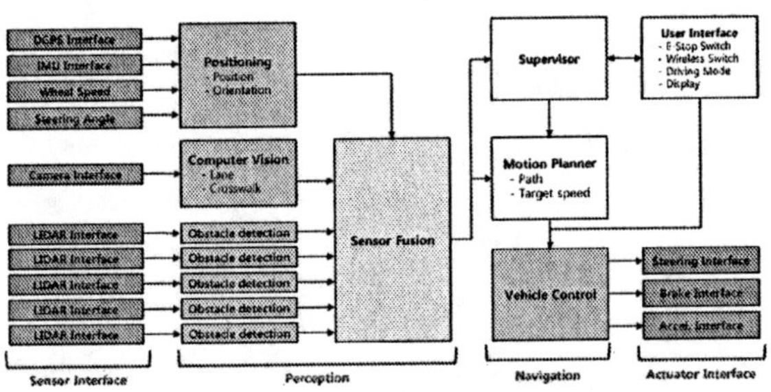

Figure 3. Software architecture of A1

Sensor Interface

The sensor interfaces manage communications with the perception software module and individual sensor. In order to utilize individual sensor data, the interface software for each sensor was implemented to the dedicated hardware node in Fig. 3.1. DGPS interface receives GPS data described in NMEA protocol from the DGPS receiver at 10 Hz, and parses useful data to estimate the vehicle position and vehicle speed [12]. IMU data are updated from IMU sensor at 100 Hz, and the steering and four wheel speed data are transmitted through the IVN at 100 Hz. These data are transferred to the location algorithm in the perception part. Camera interface provides the color image to the vision module in the perception group. LIDAR interface communicates with the laser radar sensor and provides range data of obstacles to the obstacle detection module.

Perception

Perception consists of positioning, computer vision, obstacle detection, and sensor fusion. In the positioning module, vehicle position, orientation, and vehicle speed information are estimated. In particular, the position and orientation data for autonomous driving are required to be accurate, reliable, and continuous. However, a stand-alone GPS-based positioning system cannot fulfill these requirements due to the occasionally temporary loss of satellite connection and signal errors. In order to provide continuous, accurate, and high integrity position data, the positioning system was aided by additional sensors such as INS, in-vehicle sensors. The positioning system of A1 utilizes the Bayesian filter-based sensor fusion scheme. The performance of a localization system depends on the appropriate choice of process model for the construction of the Bayesian filter. However, it is very difficult to choose the optimal model satisfying all the driving condition. For this reason, the interacting multiple model (IMM) estimator was applied. The purpose of the computer vision system of A1 is detection of the lane and crosswalk while driving on the paved road. First of all, in order to reduce noises of the image captured from a single camera, the Gaussian filter and morphological image processing are applied. For the detection of lane marks, the edge detection and the Hough transform were applied using the filtered image. In order to improve the performance and robustness of the lane detection, the road boundary information estimated from the obstacle detection module was also used. Finally lateral offset of the vehicle from the center line of the road was computed from the detected lane marking, and transmitted to the sensor fusion. In the crosswalk detection, the pattern information of crosswalk was utilized to recognize the crosswalk. In order to detect the obstacle and road boundary, both forward-looking laser radar and downward-looking laser radar were used thanks to complementary features of each method. The forward-looking laser radar has a benefit to detect the obstacle at far distance. On the other hand, the downward-looking laser radar is suitable to detect the road boundary and roadside obstacle such as curb. This method can observe roadside as well as road obstacle by extracting line segments from the range data in polar coordinates. In addition, the roll and pitch of the vehicle relative to the scanning surface were also estimated and can be applied to other applications by using a roll and pitch compensation.

Navigation

The navigation modules determine behavior of the vehicle. The navigation group consists of supervisor, motion planner, and vehicle control. A key module is the motion planner, which consists of a path planner and a speed commander. Path planning and target speed generation have features that provide a system to automatically avoid the severity of collision by generation of the path and target speed. The path and target vehicle speed are transmitted to the vehicle control module. The vehicle control module computes low-level commands to the three actuator interfaces: Steering interface, brake interface, and acceleration interface. Navigation group also has a supervisor that manages the driving mode. The driving mode is determined by the user commands received from user interface consisting of emergency stop switch, wireless switch, driving mode interface, and display.

MOTION PLANNING

Figure 4 presents the structure of path planning and target speed generation for autonomous vehicle A1. The goal of the path planning algorithm of the A1 is the generation of a smooth path from initial vehicle configuration toward the direction of the global route. The base frame, that is horizontal axis of the curvilinear coordinate system, is constructed by a parametric cubic spline using waypoints that contains the center position of the road. To utilize the road shape from the base frame, position of the vehicle is mapped on the base frame during the localization step. And path candidates are generated by using the base frame, current vehicle speed, and orientation. For the selection of a path from path candidates, the risk of collision and the smoothness of the path are considered.

PATH PLANNING

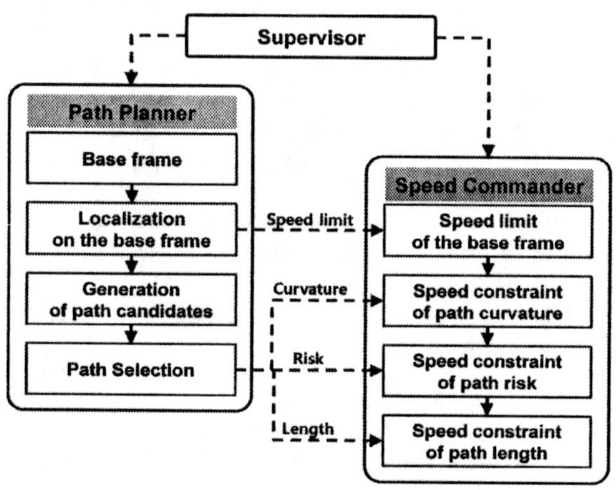

Figure 4. Path Planning and Speed Generation of A1

Localization of the Vehicle on the Base Frame

In order to design paths in the curvilinear coordinate system, mapping between Cartesian to curvilinear coordinates is required. The first step of this mapping is the computation of the closest point on the base frame to a point expressed in Cartesian coordinates. The closest point of the vehicle position on the base frame is numerically computed by combination of quadratic minimization and Newton's method which minimize the distance between the current vehicle position and a point on the base frame [13].

Generation of Path Candidates

The path planning algorithm generates a finite set of path with different lateral offsets from the base frame. Direction of path candidates is parallel to direction of the base frame [3, 4]. The resulting path satisfies the kinematic constraint of the vehicle motion. Figure 5 shows the general case for the geometric relation of path and the base frame. The arc-length of the base frame s that is the traveling distance on the base frame becomes the horizontal axis of the curvilinear coordinate system and the lateral offset q becomes the vertical axis. In order to generate a path, the function of lateral offset function is designed to provide smooth change of the lateral offset. The lateral offset q is described by the sequence of a cubic polynomial and a constant lateral offset q_c.

$$q(s) = \begin{cases} a(s-s_0)^3 + b(s-s_0)^2 + c(s-s_0) + d, & s_0 \le s < s_c \\ q_c, & s_c \le s \end{cases} \quad (1)$$

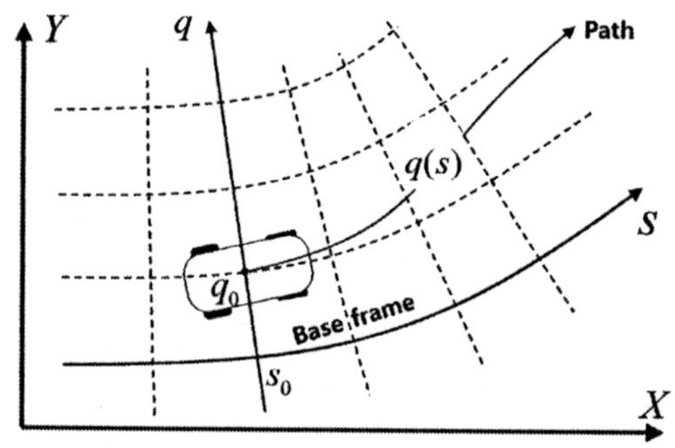

Figure 5. Relation of the path and the base frame

Path Selection

For the selection of the path, a search algorithm finds a path that minimizes a linear combination of costs that evaluate the safety, the smoothness, and the consistency of the path. It is difficult to select a safer path from the path candidates by only checking the collision for each maneuver as shown in Figure 6(a). In our approach, the risk for each path is quantitatively evaluated in blurring binary data for the collision using discrete Gaussian convolution with collision as Figure 6(b). Smoothness of a path is also important for autonomous driving, because the driving quality is mainly affected by the smoothness of path. The smoothness of a path is related to the curvature profile, which is directly linked to the lateral acceleration of the vehicle. Therefore, the integration of the squared curvature along the length of the path was adopted as the smoothness criterion of a path [14, 15, 16, 17]. Because safely and smoothness of paths are computed only using environment information which is collected at current step, excessive change of path from previous path cannot be prevented by safely and smoothness cost. Therefore, current path is compared to the previous path and its difference is considered to the selection of a path.

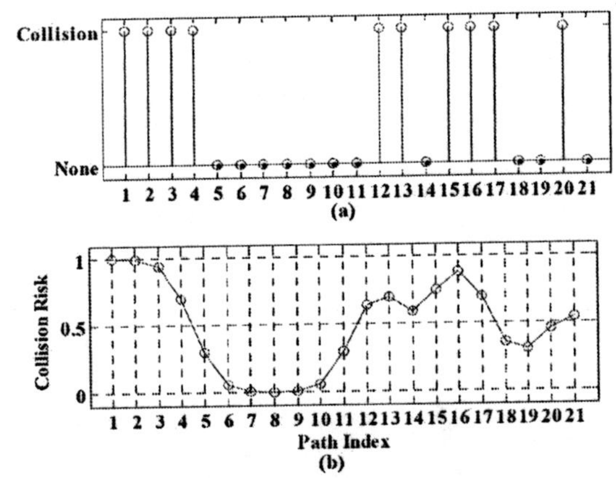

Figure 6. Distribution of collision risk of path candidates

TARGET SPEED GENERATION

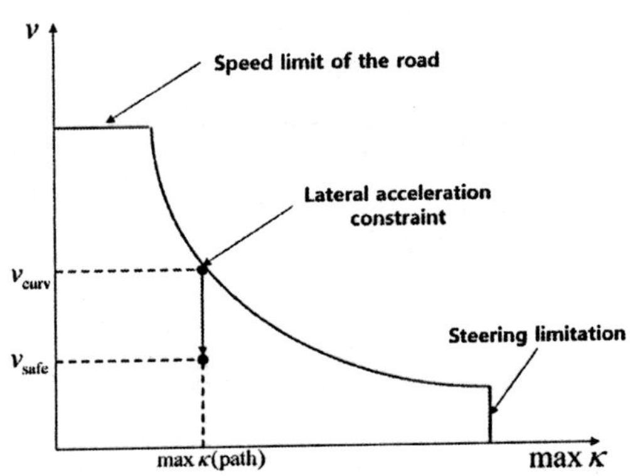

Figure 7. Speed adjustment by curvature and risk of path

The predefined speed limit of the road can be easily configured through the predefined waypoints list. However, the path planner in uncertain environments may generate the path that requires abrupt change of steering angle and huge amount of steering angle. These situations can significantly affect the tracking performance and the stability of the vehicle control. In particular, the vehicle speed and the curvature of the path are closely coupled to the lateral acceleration of vehicle motion that is directly linked to the stability. For this reason, the lateral acceleration should be limited to prevent excessive lateral forces acting on the tire. The limit of lateral acceleration $|a_y|_{max}$ is a design factor of target vehicle speed. For the conservative design of the speed limit, the maximum curvature along the generated path is used to compute speed limit. In order to drive more safely, the target vehicle speed should be determined by considering the environment information such as proximity of the obstacles and collision probabilities. The collision risk of paths provides how many obstacles exist adjacent to the selected path. For this reason, the speed commander adjusts an appropriate target speed based on the collision risk of the path as shown in Figure 7.

EXPERIMENT

On November 5, 2010, the Hyundai Motor Group held the Autonomous Vehicle Competition (AVC), which was the first race of autonomous vehicles in Korea. The goal of an autonomous vehicle in the AVC was to drive the circuit shown in Figure 8 and to complete the missions without human intervention. The winner was selected by combining the travel time and the penalties for each mission in Table 1. The driving missions were performed on both on-road and off-road. The off-road driving duties involved a tunnel, one obstacle, a gate, a median strip, a haystack, two obstacles, and a narrow road. Remainders of the missions were carried out on the on-road. The course for the 2010 AVC consisted of a 3 km of off-road and 1 km on-road. The A1 finished the course in 7 min 40 s, completing all of the missions and winning the AVC. Figure 1, 2, 3, 4, 5, 6, 7, 8, 9, 10, 11, 12, 13, 14, Figure 15 presents the logged data and snapshots obtained during the autonomous driving of the A1 at the AVC course. Figures drawn using the logged data of the AVC race illustrates the obstacle, the base frame, the vehicle position and orientation, the trajectory of the vehicle, the path candidate, and the selected path from the path candidates. The black lines in figures represent obstacle data, which were collected and filtered during the racing, and the green lines represent the base frame that was generated using the given waypoints. The dashed blue lines indicate the path candidates, while the solid blue line is being selected as a path from the path candidates. Figure 9, Figure 10, Figure 13, Figure 11, Figure 14, Figure 12, and Figure 15 present logged data for autonomous driving at tunnel mission, the one obstacle mission, the gate mission, the median strip mission, haystack mission, two obstacle mission, and narrow road mission respectively in Table 1 - Mission of Autonomous Vehicle Competition. All other missions in Table 1 were also cleared by using same manner that consists of path planning algorithm and target speed generation.

Table 1. Mission of Autonomous Vehicle Competition

Missions	Penalty
Tunnel	+3 min
One obstacle avoidance	+3 min
Gate	+3 min
Median strip	+3 min
Haystack	+3 min
Two obstacles	+3 min
Narrow road	+3 min
Lane keeping	+1 min / 20 m
Speed Limit	+5 min
Stop at crosswalk line	+5 min

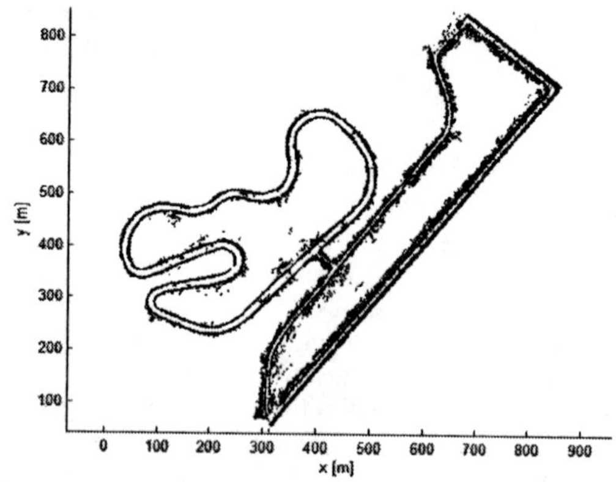

Figure 8. Accumulated obstacle map of AVC course

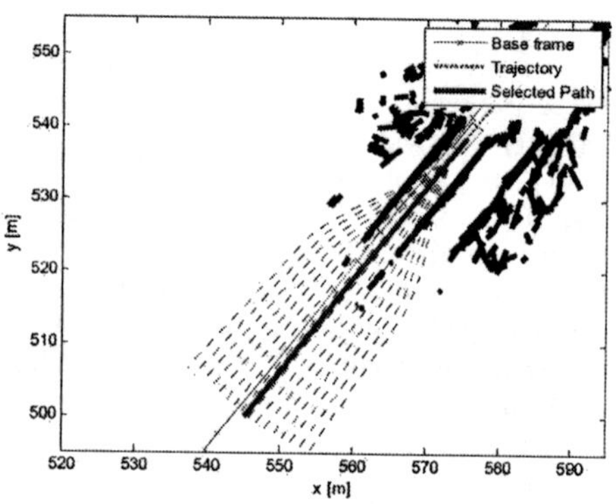

Figure 9. Tunnel

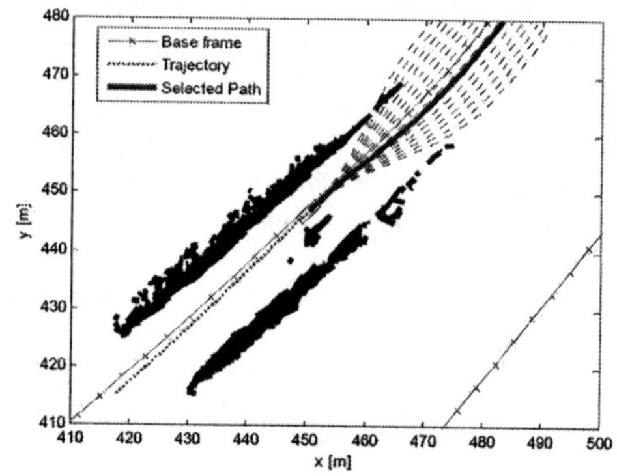

Figure 10. One obstacle

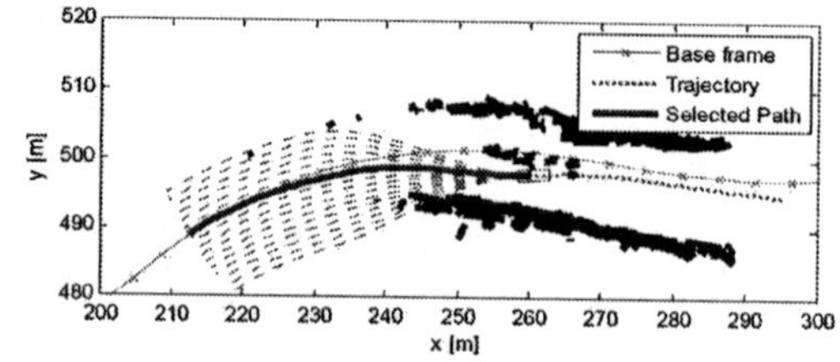

Figure 11. Median strip

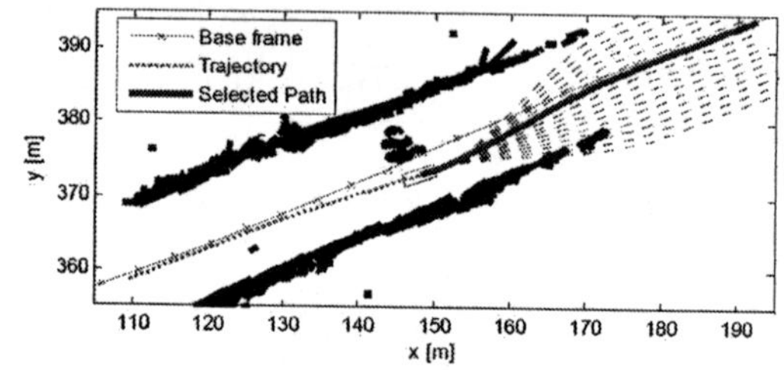

Figure 12. Haystack

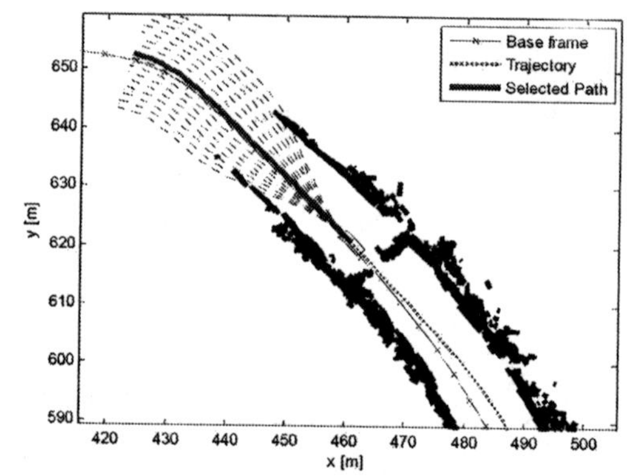

Figure 13. Gate

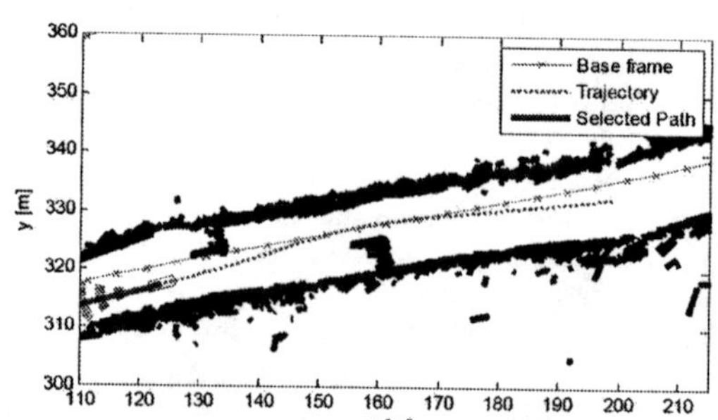

Figure 14. Two obstacles

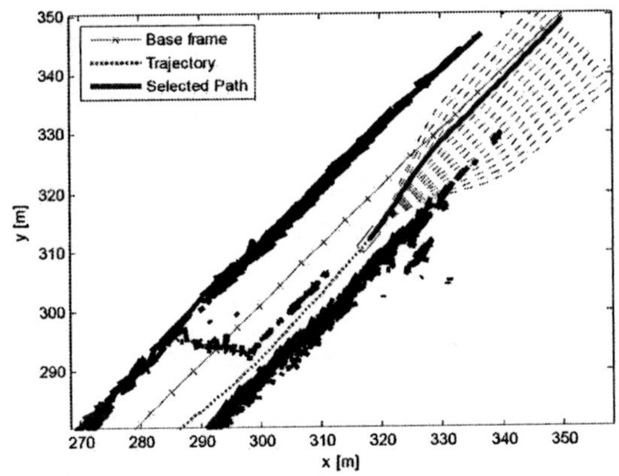

Figure 15. Narrow road at finish line

SUMMARY/CONCLUSIONS

This paper described a development of an autonomous vehicle and a planning algorithm. For the development of autonomous vehicle, network-based-system architecture was designed and implemented to the autonomous vehicle A1. Network based autonomous driving system in this paper provides a distributed computing system for autonomous driving. The distributed computing system of A1 encapsulates both logical-level function and the associated hardware into one subsystem. The software of the A1 consists of five groups which correspond to the following functions: sensor interface, perception, navigation, actuator interface, and user interface. This modular software architecture enables each software module to be individually developed and tested. Among several functions of the autonomous vehicle, this paper also presents a real-time path planning and target speed generation algorithm of autonomous vehicles based on the curvilinear coordinate system. Our strategy of path planning and target speed generation scheme provide a design framework for path planning considering safely, smoothness, and consistency of path. The proposed autonomous driving system and planning algorithms were successfully applied to the autonomous vehicle A1, which won the 2010 Autonomous Vehicle Competition.

REFERENCES

1. Bishop, R., Intelligent vehicle technology and trends. 2005: Artech House.

2. Li, L. and Wang, F. Y., Advanced motion control and sensing for intelligent vehicles. 2007: Springer.

3. Thrun, S., et al., Stanley: The robot that won the DARPA Grand Challenge. Journal of Field Robotics, 2006. 23(9): p. 661-692.

4. Ferguson, D., Howard, T.M., and Likhachev, M., Motion planning in urban environments. Journal of Field Robotics, 2008. 25(11-12): p. 939-960.

5. Montemerlo, M., et al., Junior: The Stanford entry in the urban challenge. Journal of Field Robotics, 2008. 25(9): p. 569-597.

6. Urmson, C., et al., Autonomous driving in urban environments: Boss and the urban challenge. Journal of Field Robotics, 2008. 25(8): p. 425-466.

7. Autonomous Vehicle Competition Champion A1 (driverless car). Available from: http://www.youtube.com/watch?v=AR-RcQp2a6c.

8. Autonmous Vehicle Competition.; Available from: http://avc.hyundai-ngv.com/index.php.

9. Jo, K., et al. Integration of multiple vehicle models with an IMM filter for vehicle localization. 2010.

10. Park, I. and Sunwoo, M., FlexRay network parameter optimization method for automotive applications. IEEE Transactions on Industrial Electronics, 2011. 58(4): p. 1449-1459.

11. Jang, K., et al., Design framework for FlexRay network parameter optimization. International Journal of Automotive Technology, 2011. 12(4): p. 589-597.

12. Hofmann-Wellenhof, B., Lichtenegger, H., and Wasle, E., GNSS-global navigation satellite systems: GPS, GLONASS, Galileo, and more. 2007: Springer.

13. Wang, H., Kearney, J., and Atkinson, K.. Robust and efficient computation of the closest point on a spline curve. 2002.

14. Daniel, J., et al. Energy constrained trajectory generation for ADAS. 2010.

15. Delingette, H., Hebert, M., and Ikeuchi, K.. Trajectory generation with curvature constraint based on energy minimization. 1992.

16. Kanayama, Y.J. and Hartman, B.I., Smooth local-path planning for autonomous vehicles. International Journal of Robotics Research, 1997. 16(3): p. 263-284.

17. Horn, B.K.P., CURVE OF LEAST ENERGY. ACM Transactions on Mathematical Software, 1983. 9(4): p. 441-460.

CONTACT INFORMATION

Corresponding author:

Myoungho Sunwoo
Department of Automotive Engineering, Hanyang University, Seoul Korea
Tel: +82-2-2220-0453
Fax: +82-2-2297-5495
msunwoo@hanyang.ac.kr

ACKNOWLEDGMENTS

This work was financially supported in part by the National Research Foundation of Korea (NRF) grant funded by the Korea government (MEST) (No. 2011-0017495), by the Ministry of Education, Science and Technology through the BK21 Program (201000000000173), by the Ministry of Knowledge Economy (MKE) and the Korea Institute for Advancement in Technology (KIAT) through the Workforce Development Program in Strategic Technology, by the Industrial Strategy Technology Development Program of the MKE under (No. 10039673), and by Energy Resource R&D program(2006ETR11P091C) under the MKE, Republic of Korea.

DEFINITIONS/ABBREVIATIONS

κ
: Curvature of path

q
: Lateral offset of a path from base frame

s
: Arc-length of the base frame

X
: X position of vehicle in the Earth-fixed coordinate system

Y
: Y position of vehicle in the Earth-fixed coordinate system

v_{curv}
: Target vehicle speed w.r.t. maximum curvature of a path

v_{safe}
: Target vehicle speed w.r.t maximum curvature and risk of a path

a_y
: Lateral acceleration of vehicle

SAE International

Development of a Semi-Autonomous System for Testing with Somnolent Drivers

2011-01-0589
Published
04/12/2011

Jaime Lopez, Jose Manuel Barrios and Mario Nombela
IDIADA Automotive Technology SA

doi:10.4271/2011-01-0589

ABSTRACT

Driving is a highly complex activity which requires the driver's full attention. Presently, the human factor is related to 90% [1] of accidents and driver distraction is one of the principle causes.

The objective of this project was to create a semi-autonomous system for testing with somnolent drivers. Our system is comprised of a GPS which checks position and velocity of the car continuously, a robot which decides with data from the CAN and GPS if the vehicle is being driven properly and an actuator on the brakes and the engine to stop the car. The robot will monitor these parameters continuously in order to take control of the car if it detects that the vehicle is not being properly driven in order to avoid a possible accident. This device will be on standby.

To do this, it was necessary to create a system which allows the driver to handle the car without any influence, but when it detects driving errors can stop the car to put it in a safe state.

The system developed in the project enables testing with drowsy drivers to be carried out when there are possibilities of loss of control of the vehicle. It can also be used to increase the safety of test track facilities or other control applications outside the automobile industry, in which safety is fundamental.

INTRODUCTION

The fatigue phenomenon involves an alteration in the levels of the driver's awareness and perception, which affects the psychomotor processes that are crucial for safe driving. One factor that is particularly dangerous is the decrease in the driver's own ability to judge whether he or she is in a condition to drive. Symptoms of fatigue include frequent yawning, low concentration, and drowsiness, slow reactions, irritability, making few corrections and driving slower, difficulty staying in lane, having micro-sleeps, etc. The symptoms vary greatly among drivers, so the search for robust algorithms is not a simple task. Sleepiness is one of the most notable effects of fatigue, though not the only manifestation. Fatigue and sleepiness are extremely dangerous because they reduce driving capabilities [2].

Among the main factors involved in the phenomenon of fatigue and sleepiness are the circadian rhythms, which are the biochemical and physiological processes that follow a 24-hour cycle by which people tend to stay awake during the day and sleep during the night. These cycles mean people have a significant reduction of their driving capabilities during night-time hours and also, but to a lesser extent, between 1 and 3 o'clock in the afternoon.

Another factor is the phenomenon of micro-sleeps, which are a series of episodes of unintentional loss of attention (from minutes to a few seconds) associated with the phenomenon of daydreaming; frequently when the person is fatigued or subjected to a monotonous task such as driving. In a micro-sleep the person is not capable of responding to external stimuli. The more a person has been subjected to sleep deprivation, the more likely they are to have micro-sleeps.

The most reliable method for detecting sleepiness is the electroencephalogram (EEG) [3], since the only part of the body that sends the final order to go to sleep is the brain. Other methods for detecting sleepiness include monitoring movement of the eyes, heart rate and reflexes; however these can be affected by other physiological processes such as hunger, boredom or simply fatigue. These methods combined together can give a certain indicator of sleepiness.

The method allows the identification of sleep states through the changes in the brain waves. The states are characterized by an increase in alpha and theta waves and a decrease in beta waves. However, their measurement is possible only through intrusive measurement techniques, such as electrodes in contact with the head of the subject, which is not practical in commercial applications in vehicles.

The second group of detectors to detect fatigue or sleepiness are related to the identification of some parameters of the eyes [4]. Through monitoring the degree of openness of the eyes (PERCLOS: PERcentage CLOSing eyes) and the rate of blinking this sensor group identifies the degree of sleepiness. When entering the phase of drowsiness, blinking becomes more frequent, and its pace slower and the degree of openness and eye movement are reduced. Such systems have the drawback of not being able to detect sleepiness in people who are wearing sunglasses, as the system needs to have a clear picture of the eyes; or micro-sleeps in people who sleep with their eyes open.

Another group of sensors focus on the micro-corrections made by the driver's path through the steering wheel [5]. When entering a state of drowsiness these micro-movements are slower and diminished. Existing studies conclude that such systems are very dependent on the characteristics of the road, the driver, etc.

A fourth group of sensors monitor changes that occur in the vital signs that are affected by the onset of drowsiness, such as heart rate, respiratory rate, body peripheral temperature and blood pressure. Existing studies are inconclusive about the possibility of predicting drowsiness with only one of these variables. Due to the complexity of driving and the large variability among subjects, it seems that the solution could be a combination of systems to increase system reliability and minimize the number of false alarms [6].

The main objective of this project is the identification, analysis and determination of a minimum number of biomedical variables and / or variables of driver behaviour (movements of the steering wheel, the vehicle position in lane, etc..) necessary to detect states of alertness unsuitable for driving because of drowsiness or fatigue, and the design and development of a prototype system for detection of drowsiness or fatigue which can identify the different states of alertness. All this is obtained from a research-oriented approach to promoting safety in the vehicle.

Up to now, there have been two different kinds of test in order to collect data from drivers:

• Tests on the road with drivers driving on monotonous and tiring roads, but without feeling sleepy As soon as they felt drowsy or tired they had orders to stop the car and take a rest.

• Test on a simulator until the driver fell asleep. The aim of those tests was to pick up data of the nonreturn moment, from which the driver is sleeping and the accident imminent. However, the data were

not reliable because the environment was not representative of real driving conditions, as the subjects did not have the "tools" to prevent sleep (they can not open the windows, use air conditioning, the radio, etc.).

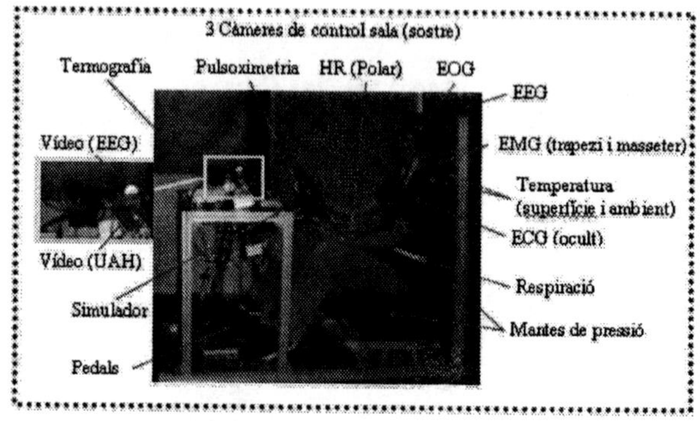

Test equipment (source FICOSA)

The road tests showed that the mental state of the drivers is not always optimal, but is still adequate for driving and to arrive at a place safely without crashing. With those data, premature alerts can be generated. Furthermore, the functionality of the data acquired can also be assessed. Analysing these signals and asking the drivers to comment on how they felt, a useful relation among the sensors was found and a first version of an algorithm developed. The most reliable data was from EEG, which, however, can not be used on a commercial system due to comfort issues.

With the simulator it was found that some drivers handled the car correctly enough, although they were asleep and did not realize it. After the test they always thought that they had not fallen asleep and lost control of the vehicle. The absence of external stimuli made the onset of sleepiness more "standard", as if they were in bed. The driver could not have elements such as radio, windows, air conditioning, etc. in the simulator, which can help to combat sleep.

The aim of these tests was for a driver to drive the car until he or she falls asleep but without accident or injury to the car or driver. All the tests must be done maintaining the safety of the driver and all the personnel involved.

The second phase of tests will be done during the night, to increase the effect of the circadian rhythms and with a driver in a state of sleep deprivation. These measures are taken to ensure the no-return point is achieved during the test. The driver is only allowed to drive on the test track, to avoid accidents before the test begins However, it could also happen that the driver does not fall sleep during the test, so it will stop when the night shift finishes.

The tests are defined as follows:

1. The circuit

Tests will be done in one of the largest dynamic platforms of IDIADA. The headquarters of the company are located 70 km south of Barcelona (Spain), having subsidiaries and branch offices worldwide countries with a total work force of more than 1.000 employees. The core services IDIADA provides are: Engineering, Proving Ground and Homologation. Main fields of engineering activity are power train, emissions, noise & vibration, vehicle dynamics, fatigue & durability and passive safety.

IDIADA's mission is to support its clients in their product development activities by providing them with engineering and homologation services that fit their needs.

The company holds the representation of the Spanish Government in several bodies dealing with automotive research and regulation, such as EEVC (European Enhanced Vehicle-safety Committee) and UN ECE (United Nations Economic Commission for Europe) Working Groups. In the same way, IDIADA is member of several working groups in which safety standards are developed (GRSP, Euro NCAP, ISO), thus permitting the acquisition of a complete view in the field of integrated safety.

The platform where the test will be done forms a cone with a circle of 300 m diameter inside. Long straights and wide curves are necessary, and the form of the circuit is very monotonous. The aim is to drive as fast as possible, and to try to get a similar feeling as on a highway, where somnolence more frequently appears.

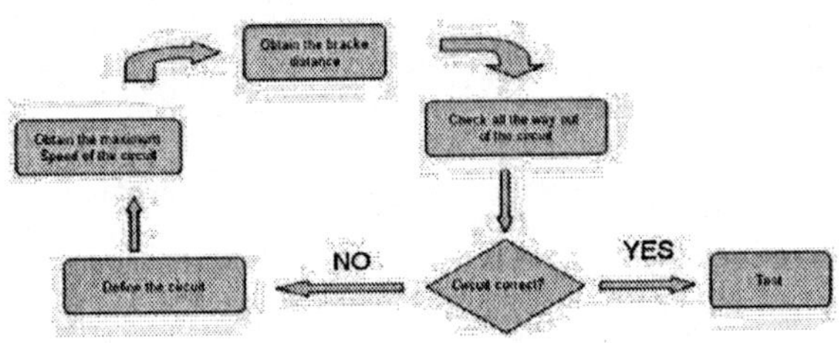

Flowchart of the test process in closed circuit

Some bends or corners to check if the capabilities of the driver are still correct or reactions too slow are also necessary. There will only be one car on the dynamic platform, namely our test vehicle.

2. The robot

A driving robot will take control at the moment the driver loses it. This robot normally controls all the parameters and the driver supervises its behaviour, but in the case of this experiment it switches roles and the robot supervises the driver with a program created especially for this test.

This program will start a thread which stops the car when the following conditions are met:

• The vehicle goes out of a predefined area. This means that the driver was not able to keep the car on the defined route. Defined by the robot brake distance, minimum distances from the perimeter of this safe area and the perimeter of the platform are established. The experiment is then finished; an accident would appear in the real world. A small lane departure is allowed before the robot starts braking.

• The vehicle exceeds a defined speed. This is measured by the GPS system and if this speed is exceeded the robot will not be able to stop the car within the platform. The driver has lost perception of the speed and this can be unsafe. The experiment must finish. No lower limit is defined. If the driver must stop the car it is possible to do so without stopping the experiment.

• The GPS location of the robot has poor accuracy. This can be unsafe and the experiment must be stopped. The robot checks the accuracy of the GPS continuously and when it is over 2 cm it sends the order to stop, as in the two conditions above.

• The computer with the program fails. In this case the watchdog square signal emitted by the computer will either be alternated (amplitude or frequency change) or fully stopped. When this is detected by the robot, the car is then stopped.

The car is stopped by actuating its braking system. The robot does this with a controlled deceleration. The robot has been set up to stop a car at 65 km/h in only 50 meters distance. The brake actuation will begin gradually and end much harder.

The robot had to be modified to be non intrusive and not disturb the diver. This modification was done especially for these tests.

An extra signal is received when the robot actuates in order to warn the team supervising the test and to launch an actuator which will stop the engine before it is stalled.

Image of the brake pedal of the robot mounted on a car (source IDIADA)

3. The environment

The entire environment must recreate the conditions of a normal road, in order to acquire the data related to driving, and the driver must be monitored. Some road markings are placed on the dynamic platform with three different intentions:

• To guide the driver on the track.

• To take data of the position of the car inside the lane, with a LDW prototype.

• To check from outside the track if the driver is doing the test correctly.

The markings must be removable and the dynamic platform ready to test with other clients in a few minutes. Besides the test conditions, all the other measurement instrumentation must work correctly during the whole experiment, because we never know when the driver will fall asleep, and these are the most important data.

4. Extra safety measures

Some extra safety measures will be taken due to the special conditions of these tests. Although the driver must be alone in the car, he will always have people taking care of his safety. A team (never a person alone) will supervise the experiment from a high point which gives a full view of the track.

The car has an emergency brake system which works directly on the calipers with oil pressure. This brake is totally independent from the robot and can only be operated by the people outside, via Wi-Fi. The way of actuating is full brake, so, the force on the driver is higher, but the brake distance smaller (about 40 meters when driving at 60 km/h). Once the brake is actuated the car is blocked and the experiment finished. The emergency brake must be reloaded manually with a special tool. The engine will also be stopped when this system actuates, since it actuates directly over the wheels of a manual transmission car.

The people outside also know when the robot is acting, so, if they receive this signal but the car does not change its behaviour, it means that a failure exists and they must stop the experiment with the emergency brake.

Before every test, a safety protocol must be followed by the personal involved in the experiment. The robot must actuate when all the four conditions are met, one by one. Finally, the emergency brake must be checked and reloaded. This procedure has two goals: the first one is to check all the systems involved and, second, to make the driver confident about all the systems mounted on the car.

CONCLUSIONS

With all these systems mounted on a single car we have created a semi-autonomous, non-intrusive system which allows free driving until the possibility of accident is detected. This makes testing possible with drowsy drivers in situations in which loss of control of the vehicle is likely.

Up to now, the initial phase of the project has tackled the gathering of drowsiness or sleepiness detection data by carrying out instrumented tests with real drivers on open roads. The results have proven satisfactory in the sense that the data collection systems (EEG, PERCLOS, micro corrections, biomedical factors) have provided enough data from the pilot tests. These data show that the selected variables are consistent with the drowsiness effects and can now be applied to the second phase of the project.

It is worth mentioning that during this phase of the project, the characteristics and time for drowsiness and sleepiness varied between the drivers selected. This will be the first time data is collected immediately before a driver falls asleep. These data are very important in calibrating the algorithm of the future somnolence detector, since other ways of obtaining the data do not give us the necessary reliability.

REREFENCES

1. Dirección general de tráfico, publication n° 150: http://www.dgt.es/revista/num150/pages/f.humano.html

Insurance Bureau of Canada, Road Safety http://www.ibc.ca/en/BeSmartBeSafe/Road_Safety/Driver_Distraction.asp

2. Siegmund, G., King, D., and Mumford, D., "Correlation of Heavy-Truck Driver Fatigue with Vehicle-Based Control Measures," SAE Technical Paper 952594, 1995, doi: 10.4271/952594.

3. Wierwille, W., Tijerina, L., Kiger, S., Rockwell, T., Lauber, E., Bittne, A.: Final report supplement - task 4: Review of workload and related research. Tech. Rep. DOT HS 808 467(4), USDOT (1996)

Liang, Yulan, Reyes, Michelle L. and Lee, John D., "Real-Time Detection of Driver Cognitive Distraction Using Support Vector Machines". IEEE Transactions on Intelligent Transportation Systems, Vol.8, N° 2, pp: 340-350, June 2007.

4. Knipling, R, Rau, P: "PERCLOS: A valid Psychophysiological Measure of Alertness AS Assessed by Psychomotor Vigilance". Federal Highway Administration Office of Motor Carriers.

5. Tijerina, L, Partner, E., Goodman, M. J. (1998). "Driver workload assessment of route guidance system destination entry while driving: A test track study". Proceedings of the 5th ITS World Congress, Seoul, Korea.

Vincent, A., Noy, I. y Laing, A. (1998). Behavioural adaptation to fatigue warning systems. Paper n° 98-S2-P-21.

Riemersma, J. B. J., Sanders, A. F., Wildervanck, C. y Gaillard, A. W. (1977). Performance decrement during prolonged night driving. Vigilance: theory, operational performance and physiological correlates. New York: Plenum Press.

Van der Hulst, M., Meijman, T., Rothengatter, T. (2001). Maintaining task set under fatigue: study of time on task effect in simulated driving. Transportation research Part F, Vol 4, pp 103-118

Lenne, M. G., Triggs, T. J., Redman, J. R. (1997). Time of day variations in driving performance. Accident Analysis and Prevention. Vol 29, pp 431-437

Thiffault, P., Bergeron, J. (2003) Monotony of road environment and driver fatigue: a simulator study. Accident Analysis and Prevention. Vol 35, pp 381-391

6. DaimlerChrysler: Driver assistant with an eye for the essentials. URL http://www.daimlerchrysler.com/dccom Last visit: january 2005

Kircher, Albert, Uddman, Marcus, Sandin, Jesper, "Vehicle control and drowsiness", Swedish National Road and Transport Research Institute. 2002.

Anon, "Perclos and eyetracking: Challenge and Opportunity". Technical Report, Applied Science Laboratories, Bedford, MA. 1999.

CONTACT INFORMATION

José Manuel Barrios
jmbarrios@idiada.com
telephone number: +34 977 166 021

Mario Nombela
mnombela@idiada.com
telephone number: +34 977 166 021

Jaime López
jlopez2@idiada.com
telephone number: +34 977 166 021

SAE International™

Investigating Control of Vision Based Autonomous Navigation in the Image Plane

2010-01-2005
Published
10/05/2010

Rachana Ashok Gupta, Wesley Snyder and W Shepherd Pitts
North Carolina State University

ABSTRACT

This paper develops a novel integrated navigational system for autonomous vehicle motion control. Vehicle control is defined in terms of the required vehicle steering angle, rate of steering change and speed. This paper proposes predictive control in the image plane. The proposed predictive control enables the navigation on the desired path, reduces the control complexity and increases the application space for multiple types of vehicles. The paper investigates vehicle control stability; especially in scenarios containing varying curvature turns and variable vehicle speeds. The primary emphasis of this paper is on vehicle control rather than scene analysis. To demonstrate the proposed vehicle control, a computer vision based multi-lane detection algorithm is introduced. The control strategy is applied such that the vehicle maintains position within the lane boundaries. Stability of the control algorithm is tested and demonstrated in multiple scenarios using 3-D simulation results.

I. INTRODUCTION

"Autonomous vehicles" are not a new concept. They potentially provide improved convenience, ease of life, and safety; both in civilian and military applications. If there is not an immediate need for for fully autonomous vehicles, there is surely for active safety assistance features.

Autonomous navigation has many aspects: determination of navigable regions, detection of other vehicles/obstacles, etc. To extend autonomous navigation to urban settings, it is required to factor in additional constraints such as detection of signs, intersections, exits, etc. In day-to-day life, humans take visual navigational capabilities for granted and many tasks (such as detecting the road/navigable regions, merging/changing lanes, following traffic rules, and most importantly, avoiding collisions, etc.) become second nature to experienced drivers. Thus, an autonomous system must have complete awareness of the entire scene in order make correct navigational decisions in urban, non-urban, or off-road dynamic scenes.

The most important component of an autonomous vehicle application is the immediate and synergistic control action. Therefore, all scene information must be converted into vehicle control parameters. The control system consists of different system components such as speed control, braking, steering motion control, etc. Taylor et al. [4] systematically compared different control strategies (lead-lag controller, full-state feedback and input-output linearization) for vision-based autonomous driving and concluded that all three strategies give comparable performance and the performance improves by curvature feed-forward component. Road parameters such as curvature are dynamically changing, and thus feedback gains in any control strategies should be changed dynamically as a function of road parameters. As we are dealing with a vision based approach, the perspective transform changes the coordinates from ground plane to image plane and thus the major challenge is to come up with a control algorithm, which will determine the correct control for every component of the integrated navigational system concurrently from image plane parameters. Graefe [1] discussed sensor fusion and multi-processor architecture for autonomous mobile robots. They suggested a highly modular vision system with a minimum intermodule communication maximizing the time available for computation. They designed some fast acting autonomous robots with the suggested architecture to prove that this hierarchical modular design is realizable. Graefe's work suggests an architecture which can be extended to a truly universal vision system for autonomous robots. Graefe, however did not address the compatibility between vision information and vehicle control explicitly. Dickmanns [5] explained a dynamic motion model for road vehicles and further used that model for the control of the autonomous vehicles in his book "Dynamic Vision for Perception and Control of Motion." The controls explained by Dickmanns are in ground co-ordinates. O'Brien et al. [3] addressed modeling and control of lateral motion of a highway vehicle and designed a steering controller to track the center of the present lane on both curved and straight highway sections using an H_∞ controller. They dealt with vehicle dynamics and external factors such as wind gust, icy conditions, etc to design the lateral controller. However they assumed the knowledge of required steering angles in ground coordinates making the control algorithm difficult to integrate with the image based lane detection algorithm.

Unlike the control algorithms presented in the literature, this paper proposes Predictive Control in Image Plane (PCIP.) PCIP determines the control for the vehicle depending on the road parameters in the image plane (considering a forward looking camera) without the transformation to the ground plane co-ordinates making it more flexible to integrate with image based road/lane detection algorithms. The architecture of the integrated navigation system presented in this paper is inspired by Graefe's work such that speed, steering and other motion components of the vehicle can be controlled independently. While these systems are controlled independently, they act as an integrated system, which realizes a stable motion control.

This paper explains the design for the controller mathematically based on only the parameters in the image plane co-ordinates determined by our vision-based multi-lane detection algorithm [6]. The controller design suggested in this paper determines the steering angle, rate of steering change and speed for the autonomous vehicle simultaneously. Additionally, the design takes into consideration that the control must be robust and stable to achieve reliable autonomous operation for the safety of the vehicle, pedestrians, cargo, drivers, passengers, etc. This paper also investigates vehicle control stability proving that predictive control based on image plane parameters is realizable.

This paper is structured as follows: Computer vision based multi-lane detection algorithm is introduced in section II. This algorithm thus decides the road/lane to follow. Mathematical foundation for the proposed control strategy in image plane for the vehicle maintains position within the lane boundaries is explained in Section III. Section IV is about stability analysis to include all of the vehicle motion components together (speed, braking, and steering system). Simulation results and stability analysis of the control algorithm is also part of Section IV demonstrating that the vehicle control is robust to on-road and off-road driving scenarios, and thus can be extended to applications in urban lane following and off-road navigation in military operations. This is followed by the conclusion in Section V.

II. VISION BASED MULTI-LANE DETECTION

Fig. 1 illustrates the projective geometry of the imaging system and result in the relationship between ground and image plane coordinates: Eq(1) and (2). Parameters are illustrated in Fig. 1. This figure is a modification of the camera configuration figure in [6].

$$y = fk_y \frac{(Z \sin\Theta - H \cos\Theta)}{(Z \cos\Theta + H \sin\Theta)} \quad (1)$$

$$x = fk_x \frac{X}{(Z \cos\Theta + H \sin\Theta)} \quad (2)$$

Where, k_x and k_y are the camera resolution in pixels / meter in x and y direction respectively.

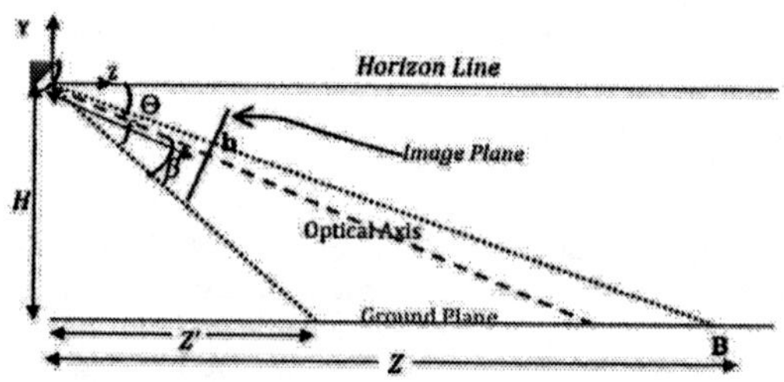

Fig. 1. Camera configuration

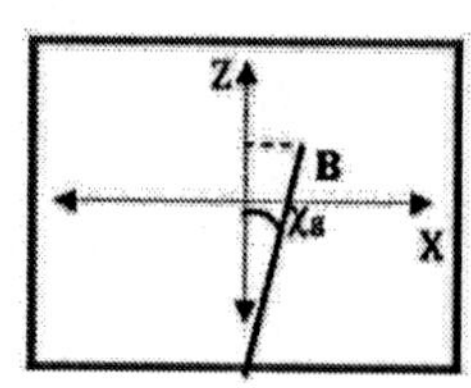

(a). Top view of the ground plane.

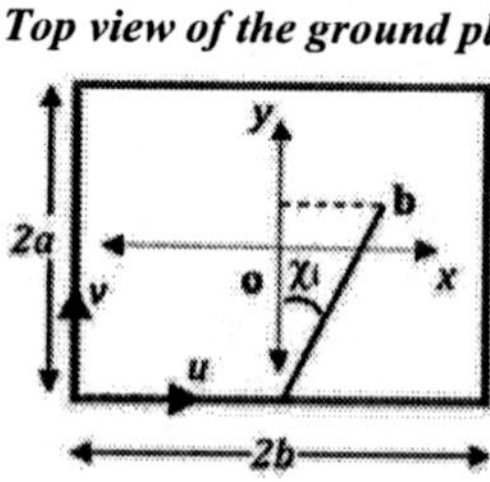

(b). Image Plane

Fig. 2. Image Plane to Ground Plane.

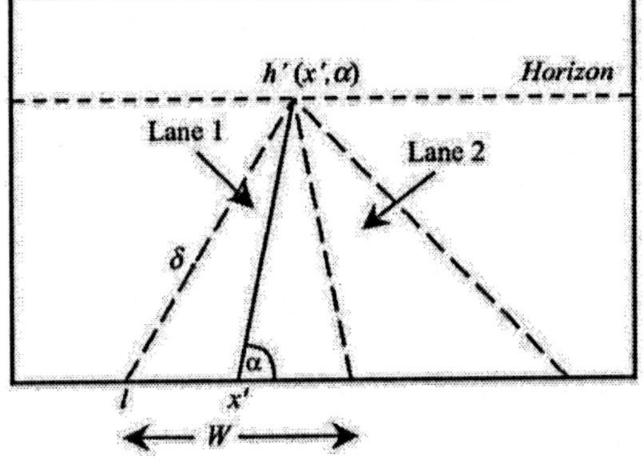

Fig. 3. Figure showing the possibility that two lanes are visible in an image. Lane edges are drawn with dotted lines. The left lane is parameterized by {x', α, W} for explanation purpose.

The first step in visual control is detecting lanes with parameters useful for control. The algorithm detects multiple lanes in the scene using the 3-D accumulator A defined in Eq(3),

$$A(x', \alpha, W) = \sum_{\delta \in \partial} \sum_{\alpha_{min}}^{\alpha_{max}} \sum_{x_{min}}^{x_{max}} p(\delta) \cdot f(W_\delta, x'_\delta, \alpha_\delta) \tag{3}$$

Where, $p(\delta)$: Weight representing the confidence that δ is a lane edge ($0 \leq p(\cdot) \leq 1$). (x', α, W) : Lane center, lane angle and lane width respectively, as illustrated in Fig. 3. Loop over all values of x', α, computing W, and incrementing the point $A(x', \alpha, W)$. Thus every edge point δ in the scene votes in the accumulator A, with a value $p(\delta)$, at the 3-D location $(x'_\delta, \alpha_\delta, W_\delta)$ representing a lane whose edges it can potentially be part of, If a lane with a parameter set (x', α, W), then all the points on that lane edge will consistently vote at the same location (x', α, W) in A forming a peak as an evidence of the lane. More details on the vision based multi-lane detection algorithm can be found in [6].

III. PREDICTIVE CONTROL IN IMAGE PLANE (PCIP)

a - Center of the bottom-most row.

p - The current location of the vehicle, (**p** is the projection of **a**, assuming that the camera is installed at the center of the vehicle.)

b(x, y) - Current look-ahead point i.e. the point to be reached by the vehicle.

c - Prediction of next look-ahead point. (x', y') - lane center. Lane center can be different from the center of the bottom row of the scene as the vehicle need not always be at the center of the lane, α - The lane angle, which the lane center makes looking at the current vanishing point.

θ - The angle made by the current vehicle position towards **b** i.e. the required steering angle to track the center of the desired lane from the current vehicle position.

l - Current look-ahead distance i.e. distance in y-direction from the current vehicle position to point **b** in image plane.

Thus, Let,

$\omega_i = \frac{d\theta}{dt}$ - The rate of change of steering angle in the image plane.

$\dot{y}$ - Vehicle speed in image plane.

We hypothesize a point **b** in the image plane towards which the car should move (Fig. 4), where **p** is the current position of the vehicle front, **b** is provided by the lane following algorithm described in [6]. A method is required to choose the steering angle (θ), the rate of change of steering angle $(\omega = \frac{d\theta}{dt})$ and vehicle speed $\dot{Z}$ as a functions of **b** and the sampling rate (dt). **b** is parameterized by the lane detection algorithm in [6] in terms of l and (x', α) shown in Fig. 4.

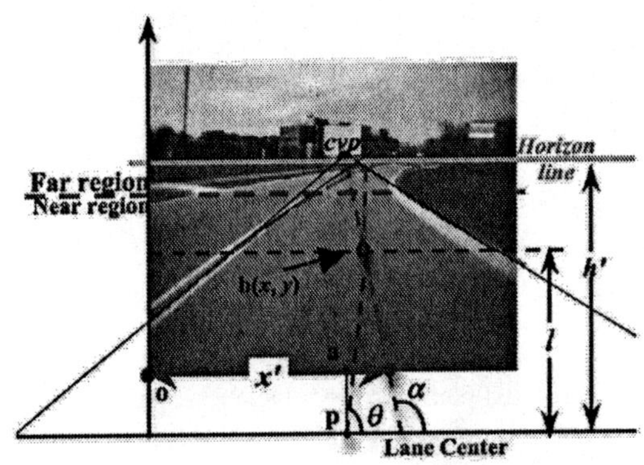

Fig. 4. Urban road scene explaining predictive motion strategy

Other than the fact that there are speed limits in the urban environment, vehicle speed is intuitively controlled by either the speed of another vehicle in front, the bending of the road and sometimes the environment and the road conditions such as icy, rainy, snowy, etc. For the vehicle to follow the road, the vehicle should always point towards the current look-ahead point **b**, assuming there are no additional vehicles. θ is then corrected by using **b**(x, y). The near region in the image plane has a point or a series of points predicting the next position(s) of the vehicle. ω can be estimated by looking at these multiple look-ahead points next to **b** if available.

A. Image plane integrated control

This section addresses the relationship between the road plane and its projective transform, the image plane, and how this affects the determination of the vehicle speed, and steering control.

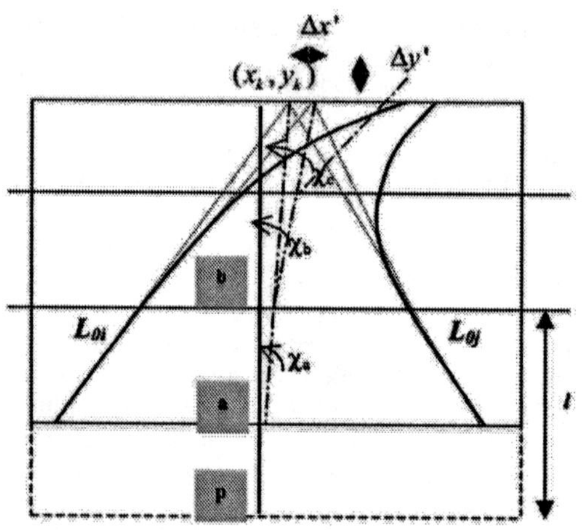

Fig. 5. Road tracking control

First the vision based algorithm is used to determine the potential road region boundaries and thus find possible road parameters in each window from bottom to horizon. As shown in Fig. 5, while at point **p**, the vehicle driver approximately predicts the next point on the road to reach at reasonably enough distance. Here as the image plane is split into multiple overlapping windows of pixel size l, we can say that the next points to reach are **a** and **b** (Fig. 5). A human driver would change the steering angle to match the new heading for the vehicle at point **a**, χ_a. In this "predictive control" algorithm, the controller approximately predicts the heading at the next point for the vehicle to reach and adjusts the speed and the steering to reach that point with the required heading.

The road curve parameters are determined from frame to frame considering the x- shift and the shift in α from the bottom-most window to the uppermost window. If L_{0i} and L_{0j} define the possible boundaries of the road region in the bottommost window, say (w_0), the current vanishing point (x_0, y_0) is the intersection of L_{0i} and L_{0j}. The change in the x co-ordinate of the vanishing point is a function of road curvature in the image plane. Considering that the road profile in the image plane follows a clothoid trajectory, the following Eq(4) holds as explained in [2].

$$C = C_0 + l\frac{dC}{dl} = C_0 + C_1 l$$
Where, $C_1 = 1/d^2, d$ is the clothoid paramter
$$\chi = \chi_0 + \int_0^l C(\tau)d\tau$$
$$\Delta\chi = \chi - \chi_0 = C_0 l + C_1 l^2/2$$

(4)

Where,

C : The curvature of the road boundary i.e. rate of change of tangent angle between two regions.

C_0 : The initial curvature of the road.

$\chi_0 = \chi_p$: vehicle heading with respect to the road axis at the current vehicle position.

χ : angle of the road boundary with negative y-axis i.e. at the bottom of the window w_0

C_1 : Rate of change of curvature.

$l/2$: the distance between the lowest rows of windows w_0 and w_1 in the image plane.

As it is assumed that l - the height of each window - is small enough to approximate that part with a straight line, the discrete form of the clothoid is considered here. Assuming the camera is installed at the center of the vehicle with a pan angle of zero with respect to vehicle heading, the current heading, χ_0 is considered to be zero. The same reason applies for projecting the current vehicle position, **p**, from the center of the scene, **a**. The steering change required at point **p** depends on the difference between the current vehicle heading (χ_P) and the required vehicle heading (χ_a).

$$C_0 = 0$$
$$C = C_1 l$$
$$\Delta\chi = \chi - \chi_0 = C_1 l^2/2$$
$$\therefore C = \frac{2\Delta\chi}{l}, C_1 = \frac{2\Delta\chi}{l^2}$$

(5)

Eq (5) estimates the curvature from point **a** to point **b** as shown in Fig. 5. The same equation can be used to predict the approximate window size l. As l should be sufficiently small to be able to approximate the boundaries as small discrete tangent line segments, Eq(6) and (7) are used as constraints in addition to determine l. The following constraints are used to track the road boundaries from the bottommost window, w_0, to the topmost window, w_3:

$$\Delta y' = |y_0 - y_1| \approx 0$$

(6)

$$\Delta x' \approx y_0 (\tan\chi - \tan\chi_0)$$

(7)

Eq(6) means the y-shift in the consecutive vanishing points from frame to frame should be negligible as we assumed that the road is flat. Where, $\Delta x'$ is the horizontal distance between two consecutive vanishing points. $\Delta x'$ is thus a function of change of heading angles (Eq(11)). The vehicle might not always be at the center of the lane and the control algorithm should be designed in a way to be able to track the lane center irrespective of the initial position of the vehicle. If the vehicle is tracking the center of the lane correctly then $\theta = \alpha$. Therefore, the required steering in the image plane, θ, can be calculated using the lane detection output (x', α) by Eq(8). (Refer Fig. 2 and 4.)

$$\theta = \tan^{-1}\left(\frac{l\tan\alpha}{l+(x'-b)\tan\alpha}\right) \tag{8}$$

Where, $2b$, $2a$: Width and height of the image plane in pixels, respectively. To control the vehicle smoothly, the steering angle needs to be refreshed based on an optimal rate of change within a given distance of the look-ahead, which is $l/2$ in this case. (See Fig. 5.)

From Eqs(2) and (1) we have,

$$Z = H\left(\frac{fk_y\cos\Theta + y\sin\Theta}{fk_y\sin\Theta - y\cos\Theta}\right) \tag{9}$$

$$X = H\left(\frac{x}{fk_y\sin\Theta - y\cos\Theta}\right) \tag{10}$$

Let the actual heading of the vehicle in the ground plane - with respect to *Z-axis* - be χ_g, which corresponds to the vehicle heading with respect to the *y*-axis in the image plane - χ_i. Z' is the ground plane distance from camera location to the closest ground point visible in the camera scene (Fig. 1): we have,

$$\begin{aligned}\chi_g &= \tan^{-1}\left(\tfrac{X}{Z-Z'}\right)\\ \chi_i &= \tan^{-1}\left(\tfrac{x}{y+a}\right)\end{aligned} \tag{11}$$

$$\begin{gathered}\lim_{X\to 0} x \to 0\\ \therefore \lim_{\chi_g\to 0} \chi_i \to 0\\ \lim_{Z\to Z'} y \to -a\\ \therefore \lim_{\chi_g\to\pm\pi/2} \chi_i \to \pm\pi/2\\ \therefore \chi_g \propto \chi\end{gathered} \tag{12}$$

Therefore left turns in ground plane are preserved as left turns in image plane and the similar is true for right turns. This makes steering control in the image plane possible without the need to transform to ground co-ordinates.

Let $\dot{y}$ be the vehicle speed in the image plane (pixels/second) and $\dot{Z}$ be the speed in the ground plane(m/s). Therefore differentiating Eq(1) and (9),

$$\dot{Z} = \frac{Hf\dot{y}}{(fk_y\sin\Theta - y\cos\Theta)^2} \tag{13}$$

$$\text{Let } g(fK_y, H, y, \Theta) = \frac{Hf}{(fk_y\sin\Theta - y\cos\Theta)^2} \tag{14}$$

Now, $\omega_i = \frac{d\theta}{dt}$ be the rate of change of steering angle in image plane. $\theta = \frac{d\chi}{dt}$ be the steering i.e. rate of change of vehicle heading. Therefore, the following estimates are used:

$$\theta_p = \frac{\chi_a - \chi_p}{\Delta t},\ \theta_a = \frac{\chi_b - \chi_a}{\Delta t} \tag{15}$$

$$\omega_p = \frac{\theta_a - \theta_p}{\Delta t} \tag{16}$$

Where, θ_p and ω_p are the steering angle and rate of change of steering angle to be applied at **p** to track the lane correctly. Let, the change in heading required from point **p** to point **a** be $\Delta\chi_p$.

$$\therefore \Delta\chi_p = \chi_a - \chi_p \quad , \quad \Delta\chi_a = \chi_b - \chi_a \tag{17}$$

$$\therefore \omega_p = \frac{\Delta\chi_a - \Delta\chi_p}{\Delta t_a} \tag{18}$$

$$\text{Similarly, } \omega_a = \frac{\Delta\chi_b - \Delta\chi_a}{\Delta t_a} \tag{19}$$

Where the speed in the image plane from point **p** to **a** is given by

$$\dot{y} = \frac{l}{2\Delta t_a} \text{ as, } y = \frac{l}{2}$$

(20)

As per Eq(13),

$$\dot{Z} = Hf \frac{l}{\Delta t_a (f k_y \sin\Theta - l\cos\Theta)^2}$$

(21)

Therefore, from Eq(14),

$$\dot{Z} = \frac{\omega_i l}{(\Delta\chi - \Delta\chi_0)} g(fk_y, H, l, \Theta)$$

(22)

Thus, from Eq(22), the ground plane speed, $\dot{Z}$, and the rate of change of steering angle in image plane, ω_i, are dependent on each other. This is intuitive as we can say that when the road is bending or for higher vehicle speeds, the steering needs to be updated faster. Thus, once the safe speed $\dot{Z}$ for the vehicle at a given curvature is calculated as per vehicle dynamics, the corresponding ω_i can be calculated recursively as the vehicle moves from current point to the predicted point using Eq(13), (18), and (22). Eq(22) will hold valid for $\dot{Z}$, as long as $\Delta\chi \neq \Delta\chi_0$.

$$\text{if } \Delta\chi = \Delta\chi_0, \omega_i = 0 \therefore \dot{Z} = \infty$$

(23)

According to Eq(23), $\omega_i = 0$ i.e. no change in vehicle heading is predicted, which is true for straight roads (curvature zero). Eq(22) thus implies infinite vehicle speed ($\dot{Z} = \infty$) on a straight road. Thus infinite speed is translated to a specific maximum vehicle speed, $\dot{Z}_{max}$. $\dot{Z}_{max}$ can be determined by various factors such as speed limit specified on the road, maximum speed of the vehicle as per its dynamic constraints, frame rate, etc.

As the relations between image plane and ground plane velocities are derived in terms of image plane parameters, the vehicle control parameters can be computed from the image plane and a transformation into the ground plane is not required. Therefore this method is called "predictive control in the image plane". Moreover, the Eq(22) shows that the vehicle controls are co-dependent and can be calculated concurrently to make sure the correct vehicle response for tracking the desired lane. This illustrates the potential of PCIP for autonomous vehicle control.

IV. STABILITY ANALYSIS AND RESULTS

The vehicle stability plays a big role in vehicle control. As we are dealing with image plane co-ordinates, the image plane control parameters are to be incorporated in the vehicle control system to achieve a desired ground plane response i.e. path following. This section presents a case study to illustrate the stability control possible with PCIP.

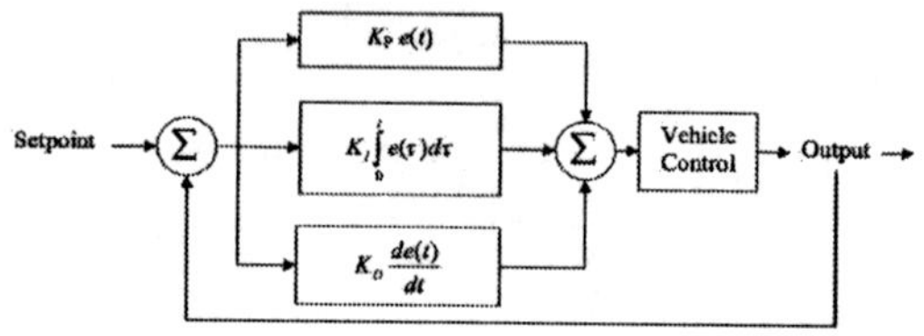

Fig. 6. Simple PID controller block diagram

A simple PID controller is employed for the vehicle steering to study the stability in lane following. K_p, K_I, and K_D are generally determined experimentally to achieve a desired response. This paper presents some initial analysis on vehicle stability. For the rate of steering angle change required, ω is calculated using Eq(22) and Eq(18). $e(t)$ is the error in the desired and current heading of the vehicle which will be used to correct the heading and thus determine the steering control. Therefore,

$$\begin{aligned} \frac{de(t)}{dt} &\approx \frac{\Delta\chi_a - \Delta\chi_p}{\Delta t} \\ \therefore \chi_a - \chi_p &= K_P(\Delta\chi_a - \Delta\chi_p) + K_D \frac{\Delta\chi_a - \Delta\chi_p}{\Delta t_a} \\ \therefore \chi_a &= K_P(\Delta\chi_a - \Delta\chi_p) + K_D \cdot \omega_p \end{aligned}$$

(24)

ω_p can be calculated using the current vehicle speed and Eq(22) for a particular value of l on a given curvature. Thus considering only PD control ($K_I = 0$), the values of required K_P and K_D, can be calculated using Eq(24). This will ensure that the vehicle response will match the speed and steering angle response required to track the road correctly. Once the PD gain values are determined approximately, Eq(24) will determine the correct ω for a required heading χa, which is determined by the image based lane detection. Further, Eq(22 can be used to determine the correct vehicle speed at the calculated ω. The gain values can be changed dynamically in case both the desired ω and $\dot{Z}$ are known in any situations. This can be determined from vehicle dynamics. In other words, if the vehicle is moving slower, the steering angle will be changed slowly by choosing appropriate PD controller gain values.

A. Simulation results

Fig. 7. Road scene rendered image from Blender

A 3-D simulation tool, Blender, is used to construct a road scene with the camera positioned to look at the road with $\theta = 10°$. A typical rendered scene image from Blender camera is shown in Fig. 7. In this simulation, the vehicle is assumed to be a point mass object. Fig. 8 and 9 show the results of the vehicle kinematics when it is operated with fixed gains, fixed vehicle speed $\dot{Z}$ (30 m/h) and fixed ω at different sampling rates without PCIP.

The figures show that as the sampling rate is decreased (increase Δt) the system becomes unstable. This results in the inability to properly track the path at $\Delta t = 0.6$ seconds due to the high curvature point. Additionally, high oscillations can be noticed near other high curvature points. In all the figures, time axis is in seconds, x and y axes are in meters. Now, as $\dot{Z}$ and ω

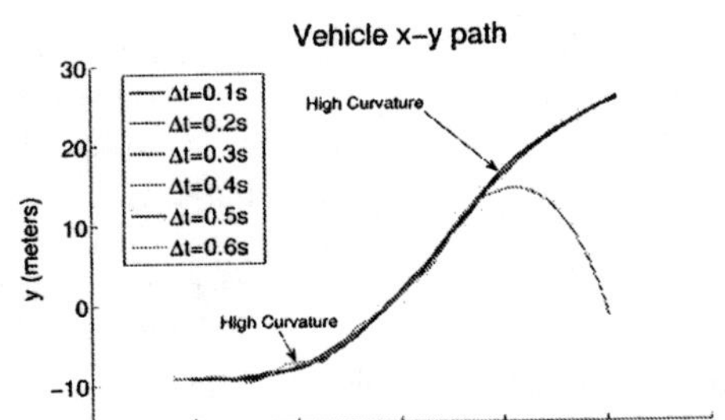

Fig. 8. Vehicle path at different sampling periods Δt without PCIP

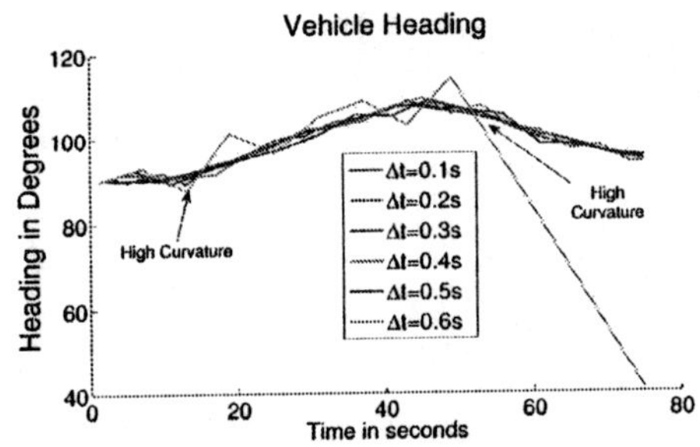

Fig. 9. Corresponding vehicle heading at different sampling periods Δt without PCIP

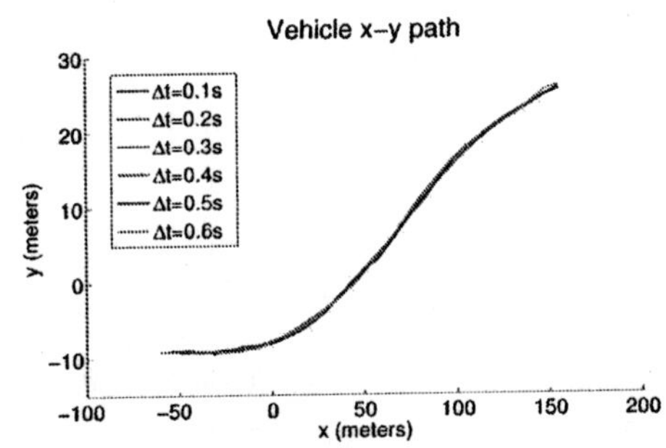

Fig. 10. Vehicle path at different sampling periods Δt with PCIP

are changed dynamically - using Eq(18), (22) and (24) - it can be observed, from Fig. 10 and 11, that stability can be achieved even for high curvature points. This is achieved by varying the control gain, speed and steering rate in accordance with sampling rate and path curvature.

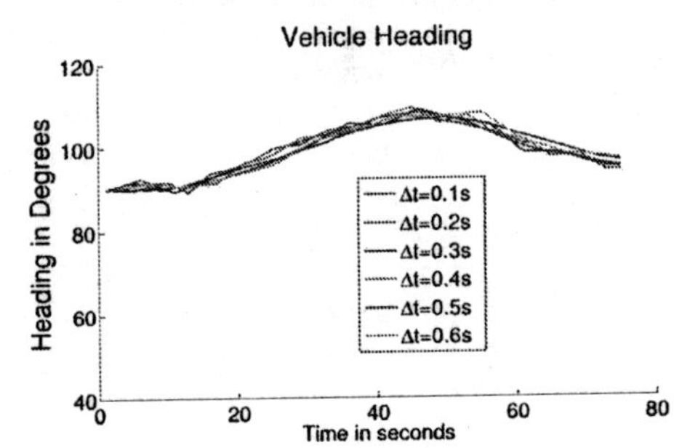

Fig. 11. Corresponding vehicle heading at different sampling periods Δt with PCIP

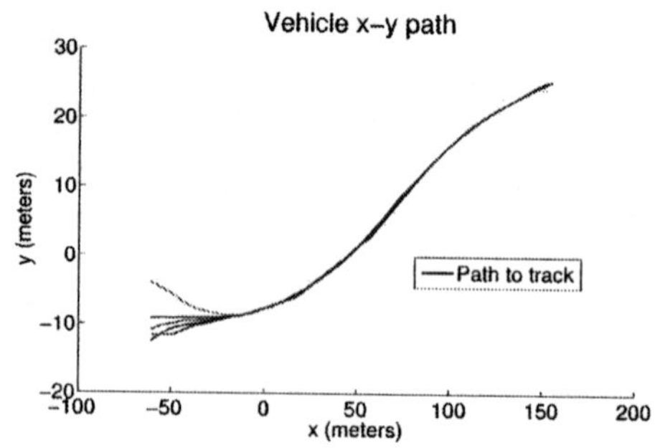

Fig. 12. Vehicle position in path tracking with different starting position with PCIP

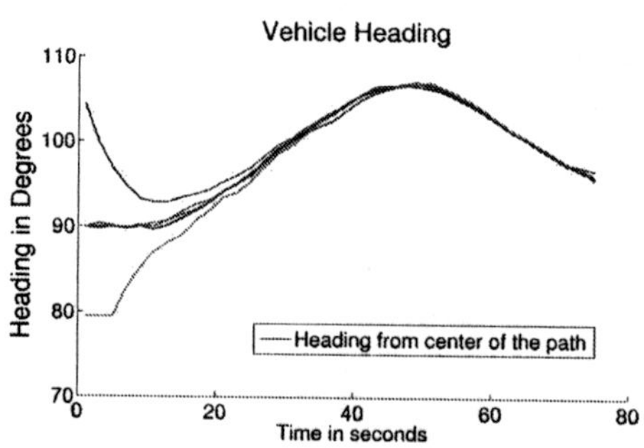

Fig. 13. Vehicle heading in path tracking with different initial heading offset with PCIP

The vehicle could go as fast as 60 miles/hour (27 m/s) with sampling period of 0.1 seconds and average speed was 25 miles / hours with sampling period as large as 0.6 seconds.

The initial condition (vehicle heading and the vehicle position) is then changed to originate off of the road. Fig. 12 and 13 illustrate that PCIP is inherently stable and tracks the path correctly, even for scenarios with aforementioned instability. Thus demonstrating that the vehicle control in image plane not only simplifies the calculations of the integrated control but is also robust to driving scenarios which originate either on-road or off-road. This method can be further extended to applications in urban lane following, lane changing and off-road navigation such as in military operations.

V. CONCLUSION AND FUTURE WORK

The paper investigates and suggests a solution to autonomous vehicle control stability; especially, in scenarios containing varying curvature turns, vehicle speeds, and sampling rate. Each necessary control parameter - such as steering angle, rate of change of the steering angle and vehicle speed - is explained and integrated into a navigational system for autonomous vehicle motion control. The integrated navigational system consists of all of the important control parameters (which are determined concurrently) in order to achieve the desired synergetic vehicle control.

Through achieving the vehicle control in image plane coordinates (i.e. without transformation to ground plane coordinates) control complexity is reduced. This method, "Predictive Control in the Image Plane," of predictive control enables navigation on the desired path. The vehicle's control of its position within the lane boundaries are demonstrated and evaluated using a 3D simulator (Blender). Stability of the control algorithm is then tested and demonstrated in multiple scenarios; such as slow sampling rate, high curvature, varying speeds using PCIP. PCIP is key to enabling the future of vision based autonomous navigation in military, civilian, and commercial applications.

In the future, PCIP will be evaluated in other driving scenarios such as lane change maneuvers, turning at an intersection, exits and merging lanes, etc. This paper investigates and proves the possibility of image plane control in terms of vehicle kinematics. PCIP will be investigated and studied for varying vehicle dynamics.

REFERENCES

1. Graefe, V. *Dynamic Vision Systems for Autonomous Mobile Robots*. In *Proceedings of Intelligent Robots and Systems and IEEE/RSJ International Workshop on the Autonomous Mobile Robots and Its Applications.*, pp. 12-23. Sep 1989.

2. Dickmanns, E. D. and Mysliwetz, B. D. *Recursive 3-D Road and Relative Ego-State Recognition*. IEEE Transaction on Pattern Analysis and Machine Intelligence, 14(2): 199-213, 1992.

3. O'Brien, R., Iglesias, P., and Urban, T. *Vehicle lateral control for automated highway systems*. IEEE Transactions on Control Systems Technology, 4(3):266-273, May 1996.

4. Taylor, C. J., Koseck, J., Blasi, R., and Malik, J. *A Comparative Study of Vision-Based Lateral Control Strategies for Autonomous Highway Driving*. International Journal of Robotics Research, 18:442-453, 1999.

5. Dickmanns, E. D. *Dynamic Vision for Perception and Control of Motion*. London: Springer, Secaucus, NJ, USA, 2007. ISBN 1846286379.

6. Gupta, R. A., Snyder, W. E., and Pitts, W. S. *Concurrent visual multiple lane detection for autonomous vehicles*. In *Proceedings of the IEEE International Conference on Robotics and Automation*. May 2010.

CORRESPONDING AUTHOR

Rachana Ashok Gupta
ragupta@ncsu.edu

ACKNOWLEDGEMENT

The authors acknowledge Lotus Engineering Inc. USA for supporting this research. The opinions expressed are those of the authors and do not necessarily reflect the views of Lotus Engineering Inc. USA. Authors also acknowledge following sponsors for their contribution, help and/or research support: Analog Devices, Automation Direct, BWI Eagle, Classic Auto Air, Comtrol, Four Brothers Auto Air, Kinetik Audio, North Carolina Center for Automotive Research, NCSU, Powertrain Control Systems, Prosilica, 1stVision, Revware, Tamron, TRW, and Vicor. The authors also acknowledge senior design students who helped to design the testing platform for experiments.

SAE International

2012-01-0741
Published 04/16/2012

doi:10.4271/2012-01-0741
saepcelec.saejournals.org

An Autonomous and Car-Following System via DSRC Communication

Chan Wei Hsu, Ming Kuan KO, Min Huai shih and Shih Chieh Huang
Automotive Research Testing Center

ABSTRACT

Inertia navigation system is capable to backup GPS unavailability. An autonomous system integrated different sensors to offer robots or remotely operated systems a reliable navigation means. This paper addresses an autonomous design method to accomplish continuously position positioning based on GPS positioning, gear-box speed, odometry, IMU and DSRC in deck reckoning. Besides, the availability of autonomous function is verified and applied using vehicles in car following or cooperative driving. This paper also presents an inter-vehicle geocasting format to accomplish multihop and car-following in cooperative driving and data exchanged based on GPS/IMU positioning and DSRC. The car following provides warning or situation awareness to drivers based on broadcast messaging the motions of neighboring vehicles obtained by one-way communication from other vehicles. The proposed system has the advantage of omnidirectional transmitting/receiving functions of DSRC module that provides 360-degree coverage in own surveillance region. The proposed system is carried out with theoretical application and hardware integration, and furthermore the result shows navigation ability and intelligent approach applicability.

CITATION: Hsu, C., KO, M., shih, M. and Huang, S., "An Autonomous and Car-Following System via DSRC Communication," *SAE Int. J. Passeng. Cars - Electron. Electr. Syst.* 5(1):2012, doi:10.4271/2012-01-0741.

INTRODUCTION

Every year in Taiwan, about two thousands deaths within 24 hours in traffic accidents, there are about 2539 deaths per hundred thousands of people and the statistical number is very serious in the world [1]. While many different factors contribute to vehicle incidents or accidents, such as rainy day or blind spot area, driver behavior is considered as the main cause of more than 95 percent. Traffic safety, in terms of infrastructure or injuries, has been discussed and improved by government's policy. However, the numbers of deaths or injuries have remained relatively flat due to the increasing number of vehicles or fatigued driving with low attention. In the recent years, more and more people like to have a team travel in the weekend. People like to take a portable navigation device with them. It provides high accuracy position, any weather condition and has the advantage in faster positioning. Although it is easier to know own location mapped onto GIS, groups of team trip cannot be aware of others position.

The inertial navigation system (INS) can overcome this shortcoming by inertia sensors. The acceleration and spatial information can be obtained from accelerometers and gyroscopes of any moving platform. An INS is an all-weather autonomous navigation system that can provide continuous position, velocity and attitude information in real-time operation [2]. The main defect of the INS is that its mean-square navigation error increases with time and needs frequent calibration with reference signals. INS error accumulates due to inertial sensor's performance with time that long period performance of INS becomes less accurate. A vehicular unit (VU) could sense vehicle speed and heading by calculating odometry. Although lower accuracy inertial sensors might cause the integration error with time in speed and vehicle spatial motion, a VU could provide continuous speed and heading with movement through controller area network (CAN). As a result, an idea of GPS/INS and VU integrated system tries to adopt commercial GPS and inertial sensors to construct a higher reliable and better accurate navigation in lower cost platform. In order to limit INS navigation errors, the INS position information could be updated in accordance with GPS, and vehicular data information is update and enforced by VU data acquisition and computation.

In the context of ITS, wireless communication plays a fundamental role in the recent two decades. In system concept, a vehicle can broadcast its driving parameters to others over Wi-Fi technologies, like dedicated short range communication (DSRC) [3]. The choice of ad hoc network, contrary to cellular network, is more rigorous and justified by the fact that the network is organized without an infrastructure which avoids data blocking or unavailability of the network as in 2.0 to 3.5G mobile communication. To meet a higher vehicle safety, DSRC which has a wireless communication protocol in the 5.9 GHz frequency band plays an important role of vehicular system. Indeed, by communicating information in remote surveillance on possible emergencies, dangerous events can be avoided. Thus, exchanged data can be used to improve the safety and become aware of neighboring vehicles location, including speed, location and heading. In addition, IEEE has taken up the standardization of DSRC by creating IEEE 802.11p [4]. In the proposed system, 802.11p protocol had been porting into embedded system for DSRC data link layer.

This paper adopted an embedded system to construct an independent navigation platform using data fusion integration for driving navigation. Combining with the GPS/IMU integration, the vehicular signals may play a potential auxiliary support to derive another package of position and moving information to enhance the autonomous capability. The hardware has been implemented on microcontrollers and carried out verification tests. The DSRC application and collision design has been presented in ARTC. Besides, the packet and data geocasting method are also designed in the proposed system. The following content will be focused on autonomous design, car-following and its verification.

SYSTEM ARCHITECTURE

The system technology is designed with an integration of INS module and data transmitting through DSRC module. The INS module could provide a good positioning solution, and then the positioning information may display on the screen to monitor other neighboring vehicles in remote operation. The vehicular data of vehicle will be routed by CAN module and broadcast to neighboring vehicles by UDP protocol onto the internet via DSRC communication. The concept of proposed system architecture is shown in Figure 1. The test information is debugged and showed in the screen of laptop using well defined format, and the total lengths which follow CAN 2.0A is about 8 bytes with its id in different devices.

A general vehicular communication which depends on its coverage area can be classified into four categories: inter-vehicle, outer-vehicle, vehicle-to-infrastructure (V2I) and vehicle-to-vehicle (V2V). A common solution, Bluetooth, its coverage is too low to do vehicular communication as a precaution. In outer-communication, mobile communication has presented its wide coverage, high reliability of data transmission in surveillance applications. However, mobile communication has a drawback in time delay about 1.0 sec in TCP mode or 0.8 sec in UDP mode [5]. To meet a high converge, data rate and low time latency. DSRC is a good choice, and its theoretically provides up to a 1 km range and allows communications between vehicles moving up to 160 km/h [6]. It also has low latency about 50 millisecond and 8 priority levels. In DSRC software, the network protocol is based on IEEE 802.11p standard under open system interconnection (OSI) model. This layer is ported from revising 802.11a, and other layers are followed UDP/IP mode.

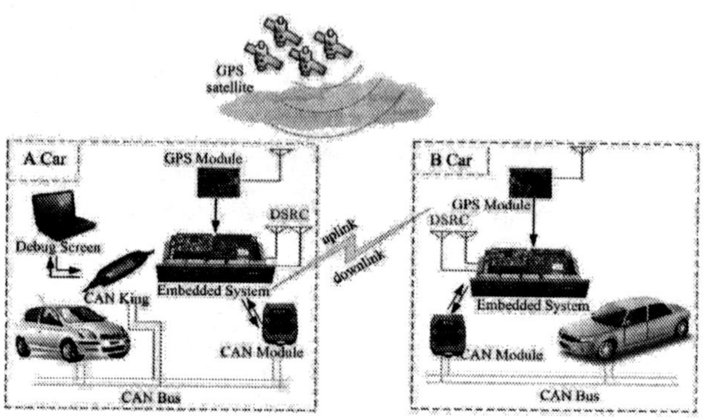

Figure 1. System architecture.

A. THE PROPOSED SYSTEM ARCHITECTURE

A data fusion of car-following system platform is built of the basic three parts, including INS module, DSRC communication unit and vehicle unit. A X86 processor is the mainly adopted controller, where the embedded kernel to access data input and output is programmed. To fulfill the proposed anti-collision application, INS data as well as vehicular data are processed in specific logic, digital formats and sent through DSRC module in controlled intervals. The data packet is formed up in X86 processor from the peripheral sensors as well as INS. Each data packet is collected and sent to DSRC module within each cycle of data surveillance. Meanwhile, the processor is activated by embedded Linux system. The USB of embedded system interface is simulated as a serial port that is used as COM1 port for INS module to capture positioning data, and two mini-PCI slots for DSRC modules to transmit data or receive data from other vehicles.

An INS module, autonomous function, is built of the basic four parts of processing core; Inertial Measurement Unit (IMU) sensors with accelerometer and gyroscope; GPS Receiver and vehicular information unit [7]. The system architecture of INS core is shown in Figure 2. The dsPIC30F6014 [dsPIC30F6014 Datasheet, Microchip Products Inc., web: http://www.microchip.com/] is chosen as the core controller to handle real time message. Time slots are used to process and measure the inertial sensors data and GPS data through Universal Asynchronous Receiver and Transmitter (UART) port.

Table 1. Requirements and specifications of INS Core.

Criterion	Required	Desirable	dsPIC30F6014
On chip nonvolatile memory	Enough to eliminate need	for data storage and use to buffer the data	144kB Flash, 8kB RAM
ADC	6 channels, at least 12-bit resolution, with 100SPS	Simultaneous sampling, ADC buffer memory	16 channels, 12-bit resolution, simultaneous S/H, up to 500k SPS
I/O pins	Enough I/O pins to support transmission	I/O pins available for more buttons	Over 48 I/O pins
Timers	At least one 16-bit timer	More timers	5*16-bit timers available
Controller Area Network interface	Data Acquisition from CAN	Vehicular data available	2 channel CAN, up to 500 kbits

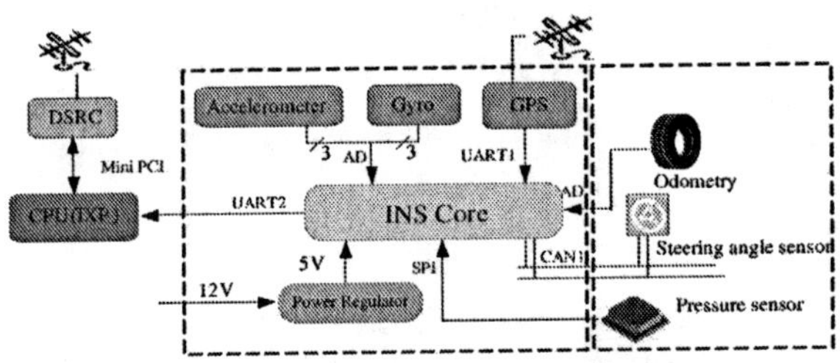

Figure 2. Block diagram of INS.

B. AUTONOMOUS CORE (INS)

The dsPIC30F6014 is Microchip product for signal processing. This chip are designed to perform as supervisor core, where GPS message and inertial analog signals are on-lined captured, sampled and processed, back to the DSRC application; while the vehicular data is determined as the calibrated information. This INS core has the communication interface to a personal computer/IXP to downlink messages and broadcast to adjacent vehicles in real time. As the needs for the system, the specification and requirement of dsPIC30F6014 are listed in the following Table 1.

An IMU is a closed system that is used to detect altitude, location, and motion. It normally uses a combination of accelerometers and gyroscopes to track the vehicle motion in attitude and location. The IMU is capable to transmit inertia data from core to IXP, where the interface is full duplex UART2. In order to output vehicle spatial states, it is accomplished by integrating an output set of sensors, such as gyros and accelerometers. Gyros measure angular rate with reference to inertial space, and accelerometers measure linear acceleration with respect to vehicle's frame. The IMU utilizes a tri-axis accelerometer and three one-axis gyroscopes as inertial measurement components. The accelerometer is measured for X-Y-Z axis; while the gyros are assigned to X-Y-Z axis correspondingly. The IMU plays a full inertial function for vehicle in real time.

The ADXRS614 operates on the principle of a resonator gyroscope. The output signal of ADXRS614 is a voltage proportional to angular rate about the axis normal to the top surface of the package. With the increase of the rotation rate, the output voltage leaves the neutral point [ADXRS614 Datasheet, Analog Devices Products Inc., UK, web: http://www.analog.com/]. An external capacitor is used to set the bandwidth. Use external capacitors in combination with on-chip resistors to create two low-pass filters to limit the bandwidth of the ADXRS614's rate response. ADXL330 is made by the principle of resonant accelerometers. It can measure both dynamic acceleration (e.g., vibration) and static acceleration (e.g., gravity). The outputs are analog voltages proportional to acceleration [ADXL330 Datasheet, Analog Devices Products Inc., UK, Rev. A., 2003. Available on web: http://www.analog.com/]. This sensor is capable of measuring both positive and negative accelerations to at least ±3 g. Because the signal measure from the accelerometer is analog, it may be disturbed by external noise. According to specifications, the external capacitor can be chosen to determine the bandwidth of the accelerometer, e.g., 0.47μF capacitor for 20 Hz is used in this paper.

C. VEHICLE UNIT FROM ON-BOARD DIAGNOSTIC CONNECTOR

The amount of electronic devices in vehicles is diagnosed by CAN bus in recent years [8]. In system platform, data is transmitted or received by CAN bus. CAN is a serial, asynchronous, multi-master communication protocol for connecting electronic control modules, sensors and actuators in automotive and industrial applications. The CAN-based system is based on the broadcast communication mechanism which is achieved by using a message oriented transmission protocol. The bit rate of CAN bus is up to 1 Mbps and is possibly operated at network lengths below 40 meter. In this study, the data rate is 500 kbps and its sampling point is held in 75%.

The CAN bus is designed and built in self-defined protocol, but the data acquisition from OBD should refer to motor standard [9]. To fulfill the proposed application, steering angle data as well as inertial data are processed in specific logic, digital formats and sent through CAN bus in controlled intervals. Each data packet is less than 8 bytes using standard ID (11 bit); moreover, the refresh time of packets is about 20ms. The CAN transceiver is the interface between the CAN protocol controller and the physical

Figure 3. System hardware and OBD II connector.

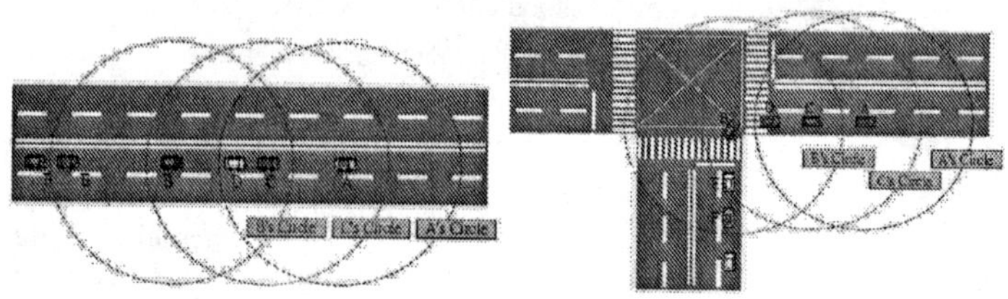

Figure 4. Inter-Vehicle Geocasting in Straight Roadway and Intersection.

transmission line and it is one of the key factors influencing the capability of network system. It is fully compatible with the "ISO 11898".

Each unit is implemented by CAN circuit board, and INS is embedded CAN controller and it adopts NXP TJA 1040 as the transceiver. The CAN transceiver is the interface between the CAN controller and the physical transmission line and it is one of the key factors influencing the capability of network system. The OBD II connector usually locates near brake/throttle under steering, and the connector is D-type and 16 pins with CAN interface. In Figure 3, the left part is system hardware and right is the OBD connector which is used to capture vehicular information.

PRINCIPLE OF SYSTEM ALGORITHM

A. INTER-VEHICLE GEOCASTING IN CAR-FOLLOWING

DSRC has the advantage of omnidirectional transmitting/receiving functions. However, it needs to depend on broadcasting messages. Geocast, i.e. the transmission of a message to some or all nodes within a geographical area, allows promising new services or application [10]. In inter-vehicle geocasting, each vehicle can broadcast own vehicular message and receive messages from neighboring vehicles [11]. In real application, each DSRC module has about 300 m of transmitting ability in this paper. This is active surveillance area, and there is another surveillance area which is about 500 m. Moreover, the outer area is relied on inter-vehicle geocasting. To meet a good geocasting, there are two key points to concern: relative distance and heading difference between prior vehicles and following ones. CCW is sent in the form of sentences; each starts with a dollar sign "#" and terminates with a carriage return <CR> and line feed <LF>. There are 9 parameters setting for CCW including group id, source node, repeater node, UTC time, latitude, longitude, height, heading, and vehicle speed. If the message is original one, the repeater node will be null string. The parameters are arranged as the sentence as follow: "#ARTC,E,, 082714,24.059958,120.383784,8.6,310.62,63.1".

In Fig. 4, left figure shows straight driving in highway or expressway and the right one is intersection case. Taking left Fig. 4 as the example, B node transmits its message periodically and receives other messages from neighboring vehicles. In B's area, it only receives messages from C, D and E. At this moment, it will determine which vehicle is located in the edge of its transmitting area. In the logical decision, B will repeat C and E message in its region if the communication time is small than 2 seconds. The communication time is calculated using Eq.(1) which is from relative position and its projection in relative coordinate. From relative position (x and y), speed (V_x and V_y) and transmitting range (R), the communication time (C_t) is obtained. For C node, it can receive A, B D and E messages from B. The previous segment is only available in straight or low curve roadway, but it cannot communicate with turned vehicle, such as right Fig. 4. In order to avoid this kind of case, the turning vehicle will broadcast vehicles message which have large heading difference relatively. Owing to non-synchronous GPS time, the message parameters include time stamp. The time stamp also uses to check the time difference and update the message by checking the effective messages.

$$C_t = (-(V_x \cdot x + V_y \cdot y) + \sqrt{(V_x^2 + V_y^2)R^2 - (V_x \cdot y + V_y \cdot x)^2}) / (V_x^2 + V_y^2) \quad (1)$$

The relative position is transformed from WGS-84 to ECEF and ECEF to NED frame using Eq.(2)-(3). The altitude (h) is given by GPS receiver and the other parameters are eccentric (e) and semi-major axis (a). Eq.(2) is result from the shape of the Earth which is an ellipsoid, not a true sphere. The following procedure is to take ownership as center and calculate relative position using Eq.(3). In Eq.(2)-(3), N is the radius of curvature in prime vertical, and two vehicle position are located in (Λ_0, λ_0) and(Λ_1, λ_1).

$$\begin{bmatrix} x^E \\ y^E \\ z^E \end{bmatrix} = \begin{bmatrix} (N+h)\cos\Lambda\cos\lambda \\ (N+h)\cos\Lambda\sin\lambda \\ [N(1-e^2)+h]\sin\Lambda \end{bmatrix} \quad N = \frac{a}{\sqrt{1-e^2\sin^2\Lambda}} \tag{2}$$

$$\begin{bmatrix} x^N \\ y^E \\ z^D \end{bmatrix} = \begin{bmatrix} -C(\lambda_0)\cdot S(\Lambda_0) & -S(\lambda_0)\cdot S(\Lambda_0) & C(\Lambda_0) \\ -\sin(\lambda_0) & \sin(\lambda_0) & 0 \\ -C(\lambda_0)C(\Lambda_0) & -S(\lambda_0)C(\Lambda_0) & -S(\Lambda_0) \end{bmatrix} \times \begin{bmatrix} x_1^E - x_0^E \\ y_1^E - y_0^E \\ z_1^E - z_0^E \end{bmatrix} \tag{3}$$

B. ATTITUDE CALCULATION - QUATERNION METHOD

Fig. 5 describes how to achieve inertial navigation via measurement and frame transform. The navigation algorithm contains several steps to compute vehicle attitude, earth rate, transport rate and Coriolis. The procedure to integrate acceleration, angular rates and calculate vehicle states in hardware is operated with software which is built in the microprocessor following the theoretical formulation below.

Euler angles are the values which present the attitude of the vehicle. The attitudes of the vehicle mean the angles between body axes with navigation axes. There are three Euler angles φ (roll), θ (pitch) and ψ (yaw) used to show the relative angles along x, y and z axis. The Direct Cosine Metrics (DCM) is used to transfer information from one coordinate system into another coordinate system. It is carried out as a sequence of three successive rotations about different axes. The DCM mentioned earlier is limited since the solution become indeterminate when θ is approximately 90 degree. The quaternion is the method to overcome this problem. The concept of the quaternion is based on the idea that a transformation from one coordinate frame to another can be effected by a single rotation angle δ and an orientation unit λ defined with respect to the reference frame in left part of Fig. 6 and right part is the definition of Euler angle.

In quaternion transformation, the orientation is written as a vector which contains four elements with the magnitude of the rotation. The preceding about body-to-navigation DCM can be expressed through quaternion elements as Eq.(4). The q_i (i=0, 1-3) are calculated from kinematic equation in Eq.(5), and it can be substituted with the quaternion elements which is shown in Eq.(6). From Eq.(6), vehicle attitude can be integrated and updated to get Euler angles.

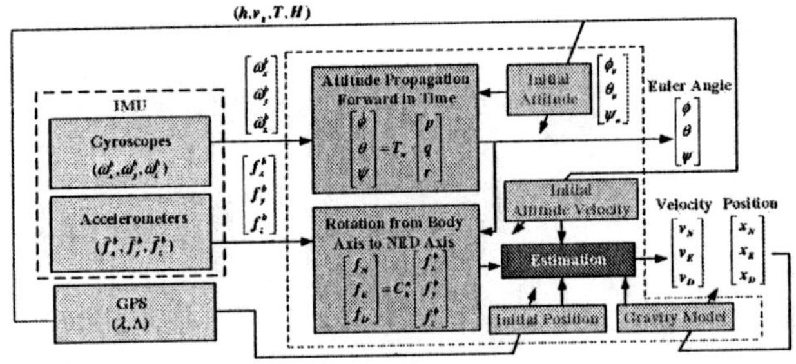

Figure 5. The block diagram of algorithm in navigation.

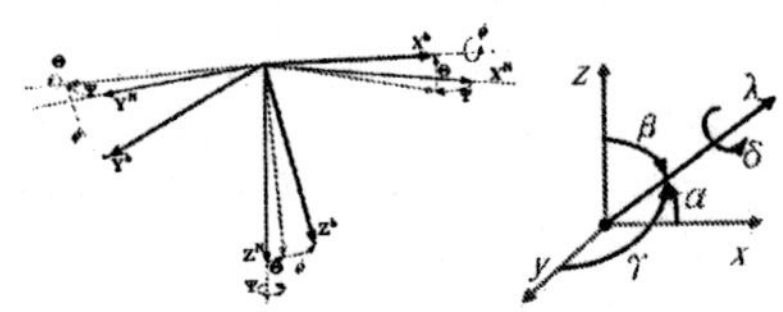

Figure 6. Definition of Euler angles and Quaternion coordinate.

$$C_b^n = \begin{bmatrix} q_0^2+q_1^2+q_2^2+q_3^2 & 2(q_1q_2-q_0q_3) & 2(q_1q_3-q_0q_2) \\ 2(q_1q_2+q_0q_3) & q_0^2-q_1^2+q_2^2-q_3^2 & 2(q_2q_3-q_0q_1) \\ 2(q_1q_3-q_0q_2) & 2(q_2q_3-q_0q_1) & q_0^2-q_1^2-q_2^2+q_3^2 \end{bmatrix} \tag{4}$$

$$\begin{bmatrix} \dot\phi \\ \dot\theta \\ \dot\varphi \end{bmatrix} = \begin{bmatrix} 1 & S\phi T\theta & C\phi T\theta \\ 0 & C\phi & -S\phi \\ 0 & S\phi/C\theta & C\phi/C\theta \end{bmatrix} \begin{bmatrix} p \\ q \\ r \end{bmatrix} \tag{5}$$

$$\begin{bmatrix} \dot q_1 \\ \dot q_2 \\ \dot q_3 \\ \dot q_0 \end{bmatrix} = \frac{1}{2} \begin{bmatrix} 0 & r & -q & p \\ -r & 0 & p & q \\ q & -p & 0 & r \\ -p & -q & -r & 0 \end{bmatrix} \begin{bmatrix} q_1 \\ q_2 \\ q_3 \\ q_0 \end{bmatrix} \tag{6}$$

The variations of velocities are integrated from the accelerations in the local geodetic frame. However, the measurements derived from sensors are in body frame. Therefore, the DCM mentioned earlier is used to transform the measurements from body frame into the local geodetic frame. The transformation is shown in Eq. (7). Owing to the Earth's rotation ($2\omega^n_{i/e}\times v^n$) and gravitation ($g^n$), the effect of the Coriolis force ($\omega^n_{e/n}\times v^n$) and gravity need to be corrected in the middle term of Eq.(7). The position of vehicle is always described with longitude, latitude, and altitude (Λ,λ,h) in local geodetic frame. The navigation systems using on earth surface are mechanized or implemented such that the local geodetic frame is maintained while the vehicle is moving. The ellipsoidal model of the Earth is used to

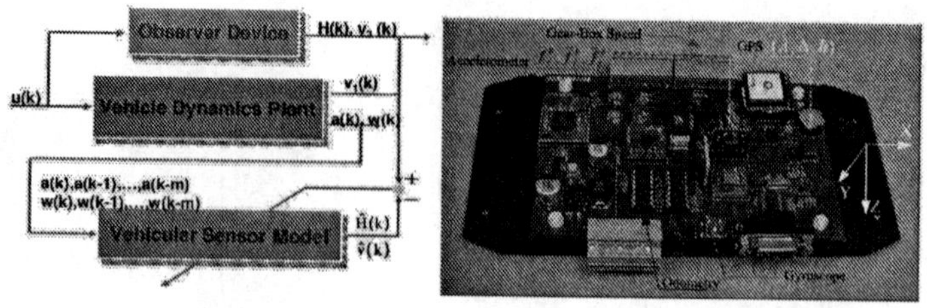

Figure 7. Inertial sensors calibration method and its hardware.

orientate the navigation frame with the variation position of the vehicle. The equations of the ellipsoidal are R_{meridian} and R_{normal}. Motion over the surface of the Earth is along the arc of the ellipsoidal surface. The changing rate of latitude and longitude are expressed in terms of R_{meridian} and R_{normal} in Eq.(8). The variation of altitude is equal to negative down velocity.

$$\dot{v}^n = f^n - (w^n_{e/n} + 2w^n_{i/e}) \times v^n + g^n \tag{7}$$

$$\begin{bmatrix} \dot{\Lambda} \\ \dot{\lambda} \\ \dot{h} \end{bmatrix} = \begin{bmatrix} \frac{1}{R_{meridian}+h} & 0 & 0 \\ 0 & \frac{1}{(R_{normal}+h)C\Lambda} & 0 \\ 0 & 0 & -1 \end{bmatrix} \begin{bmatrix} v_{north} \\ v_{east} \\ v_{down} \end{bmatrix} \tag{8}$$

C. SENSOR CORRECTION - INERTIAL SENSORS CALIBRATION METHOD

The vehicle sensing components consist of accelerometer and gyros, and furthermore they are used to perform the attitude computations in strapdown. These errors are associated with gyros and accelerometer that typically include static biases, drifts, scale factor and random noise. To solve this kind of shortcoming, inertial sensors calibration method is applied in this paper, as shown in Fig. 7. The vehicle plant is demonstrated vehicle, and the output is captured from IMU. In calibrated operation, this paper corrects gyro and accelerometer parameters. Taking gyro procedure as example, the observer device used GPS to get vehicle heading. Refer to Eq.(9)-(10), system controller integrated angle rate and compared with heading. This method uses second-order minimal energy and gradient method to get error variation in Eq.(11). The relation can be derived to discrete form in Eq.(12).

$$\hat{H}[k] = \int (\hat{p}_n w^n + \hat{p}_{n-1} w^{n-1} + \ldots + \hat{p}_0) dt \tag{9}$$

$$e_H[k] = H[k] - H[k-m], \quad e_{\hat{H}}[k] = \hat{H}[k] \tag{10}$$

$$J = \frac{1}{2}(e_H[k] - e_{\hat{H}}[k])^2 \tag{11}$$

$$\begin{bmatrix} \hat{p}_n[k] \\ \hat{p}_{n-1}[k] \\ \vdots \\ \hat{p}_0[k] \end{bmatrix} = \begin{bmatrix} \hat{p}_n[k] \\ \hat{p}_{n-1}[k] \\ \vdots \\ \hat{p}_0[k] \end{bmatrix} + \lambda \cdot (e_H[k] - e_{\hat{H}}[k]) \cdot \begin{bmatrix} w^n[k] \\ w^{n-1}[k] \\ \vdots \\ 1 \end{bmatrix} \tag{12}$$

SYSTEM IMPLEMENTATION AND TESTS

The developed system is integrated and fabricated modular units based on the circuit configuration. All the modules are available and ready to use. The implementation work needs to design suitable power supply; data interface and control circuit with accurate strategy to carry out the expected function capability. System software on data bus is programmed with appropriate protocol. Under the integrated concept, the proposed system will operate GPS data acquisition and INS correction to enhance the navigation performance. To accomplish the capability of VU, the odometry and gear-box speed are test and compared with GPS velocity in ARTC campus.

A. ODOMETRY AND GEAR-BOX SPEED TEST

The odometry hardware used frequency to voltage chip to convert signals into voltage. The dynamic test is compared with GPS velocity. The odometry speed test is implemented in the ARTC campus, and operator drove to the road terminal and turn left/right. Fig. 8 showed the variation and difference contrast to GPS speed. The frequency of odometry signal was varying from 0 Hz to 600 Hz, and then converted to voltage (0.0~5.0V) using charge-pump. In low speed operation, the odometry signal has large variation because of disturbance in Fig. 8(a)-(b). However, the mostly result is very similar to GPS velocity and this test is accomplished in calm weather. The Fig. 8(b) is special used to test availability and know how the available speed is. The odometry sensor has high feasibility and could be used in vehicle test when the speed is larger than 10kph. Although the odometry had shortcoming in low speed driving, the proposed system adopts gear-box speed as alternative solution in low speed. The gear-box speed is used to instead of odometry in low-speed, but higher speed is also adopted from odometry information because of turning angles. The gear-box signal is digital level, and its frequency varies from 0.5 Hz to hundred Hz with speed variation. Fig. 9 is the vehicle dynamic test comparing with GPS speed.

B. INERTIAL SENSOR CALIBRATION AND TEST

After INS had been set well in demonstrated vehicle, a driver drove in different speed to test straight moving and turn availability. The straight driving test is used to adjust accelerometer parameters refer to GPS speed. The output data would be processed using integration. Owing to integration error, the result should be calibrated and delicate processed well. Fig. 10(a) used a one-axis acceleration to get speed and the result is compared with GPS speed. The parameters was learned and calibrated by parameters learning and error cancelation. Hence, the speed error is under 5kph. In the similar way, the gyro integrated angular rate into heading comparing with GPS course w.r.t North direction in Fig. 10(b).

C. DISTANCE TEST IN AUTONOMOUS FUNCTION

This paper adopted a realization test in a car. In the integration of INS, the GPS provide a good position and heading. The reliability for long period is well, but it might be affect by the environment. The characteristic of INS is autonomous and reliable in short period. But the integration error is as larger as time. So the integration of GPS and IMU has a good advantage in compensating each other. The GPS can be an error bound of INS. Fig. 9 is the test area, and the related experiments are test in different driving procedures from reference points located in the terminal of the road. The test included static tests in Table 2 and dynamic result in Table 3.

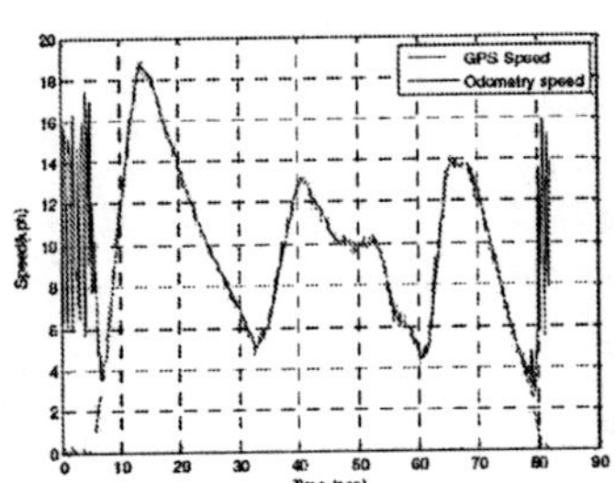

Figure 6(a). Driving test from 0~20kph.

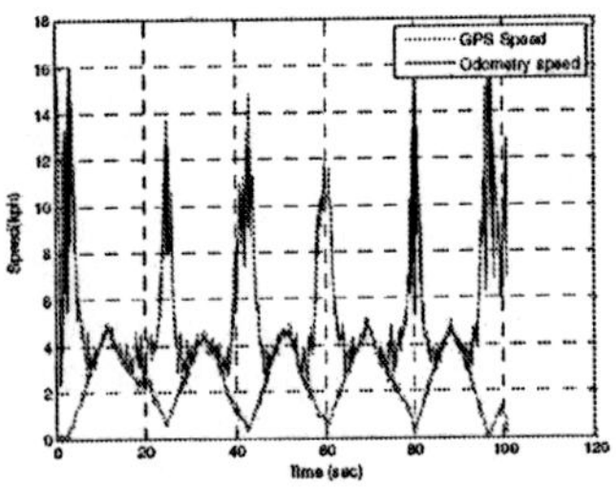

Figure 6(b). Driving test from 0~5kph.

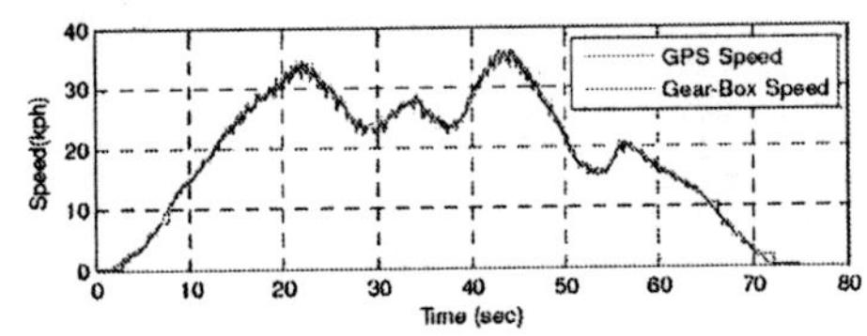

Figure 7(a). Gear-box speed vs GPS speed.

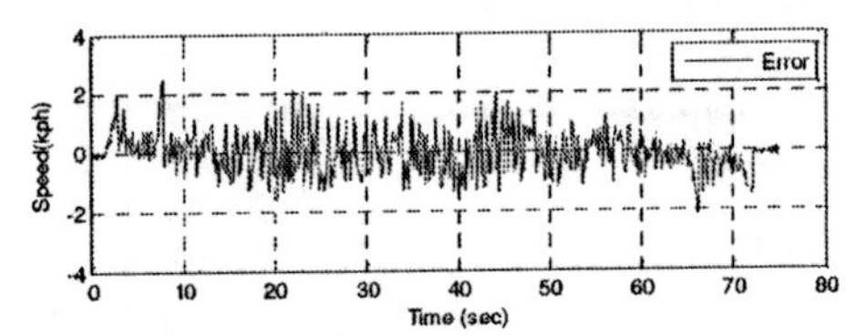

Figure 7(b). Speed error vs GPS speed.

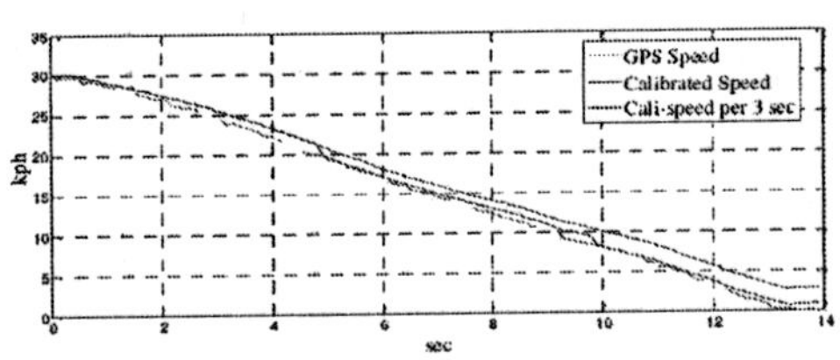

Figure 8(a). Acceleration integration test.

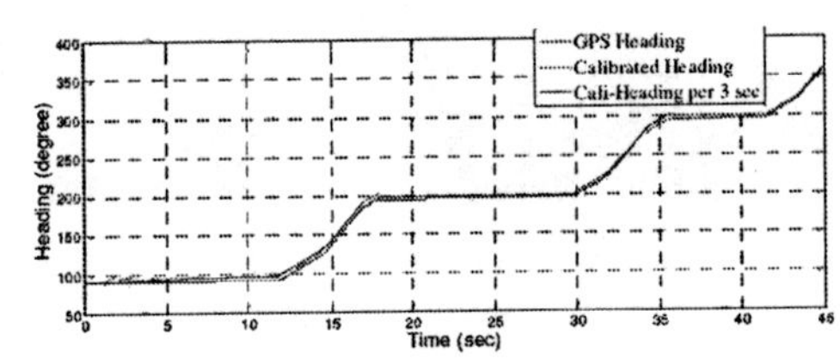

Figure 8(b). Gyro integration test.

Table 2. Static tests result.

Ref. Points / Error index	ARTC7	ARTC8	ARTC9	ARTC10	ARTC11	ARTC12	ARTC13	ARTC14
RSS (m)	10.097	7.062	7.592	6.487	5.351	4.904	8.856	4.200
	10.097	7.062	7.666	6.486	5.350	7.998	8.320	4.200
North Error(m)	-7.140	-4.994	-5.364	-4.587	-3.783	-3.468	-6.257	-2.970
	-7.139	-4.993	-5.420	-4.586	-3.783	3.952	-5.871	-2.970
East Error(m)	3.601	3.990	5.790	4.874	3.289	5.570	3.950	5.458
	3.959	3.823	5.842	4.709	4.063	-1.981	3.950	5.458
CEP50 (m)	0.1718	0.144	0.295	0.145	0.341	0.115	0.338	0.1
	0.158	0.1	0.126	0.1	0.158	6.952	0.451	0.1
CEP90 (m)	0.308	0.253	0.537	0.255	0.621	0.182	0.616	0.1
	0.280	0.1	0.212	0.1	0.280	12.674	0.821	0.1

Table 3. Dynamic tests result.

Tests / Reference points	Test1	Test2	Test3	Test4	Average distance and its error rate	Max. error rate
Ref.12→Ref.11(N to S)	97.10	97.03	97.47	95.95	96.89	
•RTK 97.8 m	0.72%	0.79%	0.34%	1.89%	0.93%	
•RMS 96.4 m	0.73%	0.65%	1.1%	0.47%	0.74%	
Ref.10→Ref.9(W to E)	224.23	223.06	224.53	223.24	223.77	
•RTK 221.9 m	1.05%	0.52%	1.19%	0.60%	0.84%	
•RMS 221.2 m	1.37%	0.84%	1.51%	0.92%	1.16%	
Ref.8→Ref.7(S to N)	93.92	95.04	92.72	92.72	98.5	3.32%
•RTK 95.9 m	2.06%	0.90%	3.32%	3.32%	2.4%	3.22%
•RMS 95.8 m	1.96%	0.79%	3.22%	3.22%	2.3%	
Ref.14→Ref.13(E to W)	224.27	223.47	223.24		223.66	
•RTK 222.7 m	0.70%	0.35%	0.24%		0.43%	
•RMS 225.8 m	0.68%	1.03%	1.13%		0.95%	

Table 4. Dynamic tests result.

Position (m)	Packet Loss Rate (%)	Latency (msec)
+200	76	654.8
+100	70	461.7
0	55	19.6
−100	66	668.6
−200	54	361.3

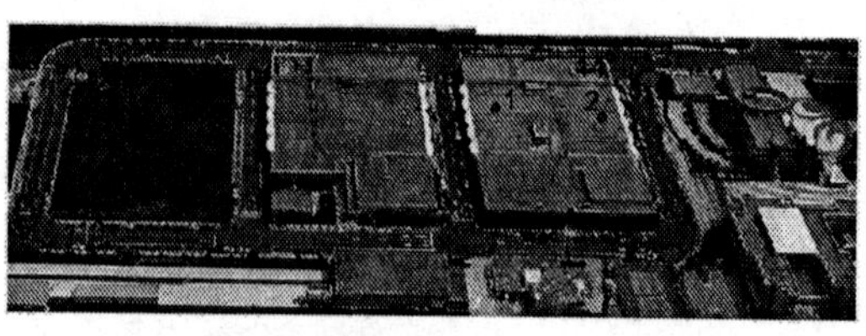

Figure 9. Test field in ARTC campus.

D. CAR-FOLLOWING INTEGRATION TEST

Under system implementation, the platform static tests include packet loss and latency. The environment noise was measured before static test in Fig. 10. The power of environment noise was −115 to −95 dbm, and it was too small to affect the packet tests. Each result would minus 20 because of probe decay. The range test showed in Table 4, and the packet loss rate rule is to get 95% successful receiving at 1000 times packet transmitting.

After the static tests were finished, the dynamic test of cooperative driving was operated. The test platform, as shown in Fig. 11, was used to implement the proposed car-following function. Three vehicles were used to run on the ARTC roadway to verify the geocasting and neighboring situation function; while three vehicles were followed on roadway or driven to an intersection in order to easily verify system design in the verification. Under system design and implementation, vehicular data which connected to CAN interface was collected for scheduled tests. For test operations, DSRC module was used to exchange positioning

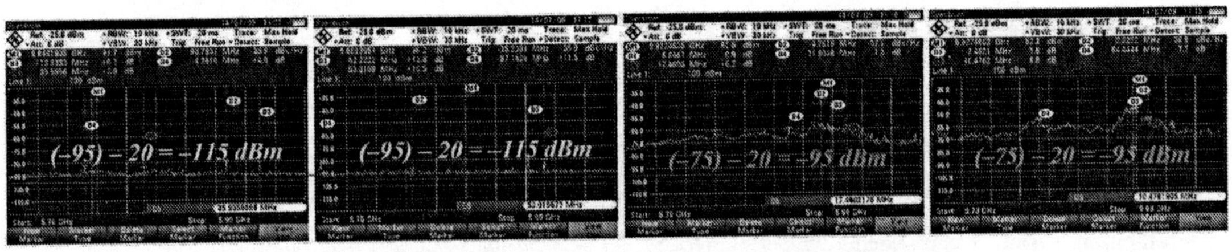

Figure 10. Background measurement using frequency spectrum.

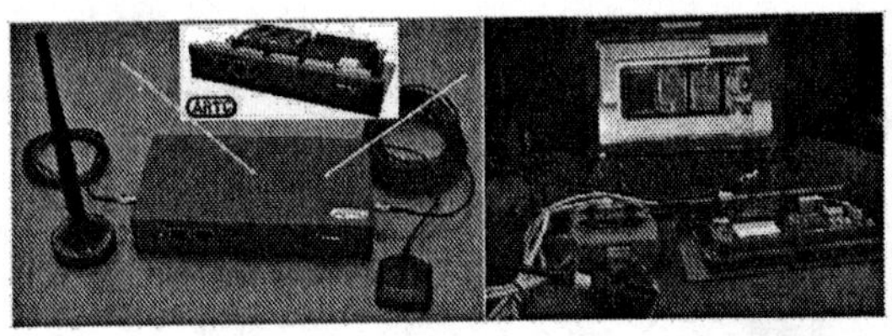

Figure 11. System Hardware and Implementation.

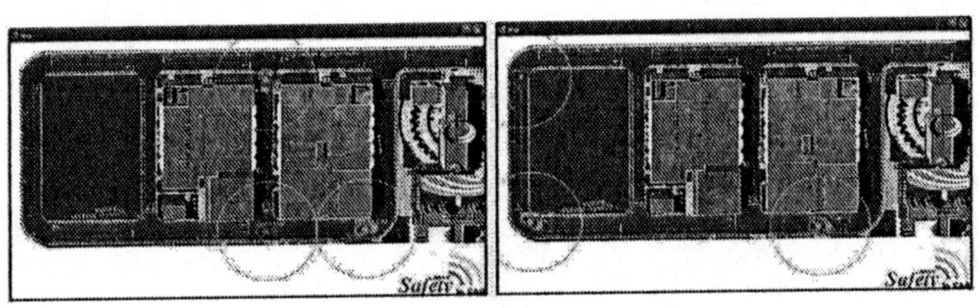

Figure 12. Inter-vehicle geocasting result in intersection

data. The INS module also outputted positioning data, and the positioning performance was well processed and mapped to demonstrated map. Fig. 12 showed the inter-vehicle geocasting tests, where two cases showed that three vehicles were individually running on ARTC roadway. Two tests demonstrated the car-following away from other vehicles, and the driver drove far away from preceding vehicle or following the other vehicle. Fig. 13 showed another test, where the driver drove into an intersection. These cases offered an important awareness to the driver under test. The map reported neighboring vehicles position periodically, and broadcasted to other vehicles via DSRC communication. In these tests, the actual position of the vehicle was monitored and displayed on the screen. The system concept provided a car following that the operator could be fully aware of 360-degree vehicles situation under system concept.

Figure 13. Car-following in blind intersection

CONCLUSIONS

In this paper, the proposed concept demonstrates a cooperative driving or car-following concept for remote surveillance applications. The system design simplified using embedded microprocessor with DSRC module to activate UDP and port 802.lip protocol into data link, continuously vehicular speed using CAN network and high sensitivity INS positioning. The proposed system assists drivers to know current relationship to other vehicles through intersection and following tests with the situation awareness capability. Although the result of DSRC packet loss rate and latency are not very good, this drawback will be improved in progress concern. The demonstrated tests have been verified the availability of collision estimation and inter-vehicle geocasting algorithm.

Several hardware, software and firmware are implemented to realize the concept of autonomous navigation. Combining with the GPS/IMU integration, the vehicular signals could play a potential auxiliary support to derive moving information to enhance capability. The advantage of autonomous system has presented fewer than 4% position error, and the demonstration provides a higher availability solution for vehicle position to enhance cooperative driving.

REFERENCES

1. "Death statistics from 1999 to 2010 of road traffic safety portal transportation in Taiwan". Available at http://168.motc.gov.tw/GIPSite/wSite/mp
2. Xiaohong, (Allison), Zhang (M.Sc. thesis), "Integration of GPS with a medium accuracy IMU for metre-level positioning," *UCGE Report No. 20178*, June 2003.
3. Sengupta, R., Rezaei, S., and Shladover, S. E. et al., "Cooperative Collision Warning Systems: Concept Definition and Experimental Implementation," *Journal of Intelligent Transportation Systems*, Volume 11, Issue:3, July 2007, pages 143-155.
4. Jiang, D., Taliwal, V., Meier, A., "Design of 5.9 GHz DSRC-Based Vehicular Safety Communication," IEEE Transactions on Wireless Communications, Oct. 2006, Vol. 13, Issue: 5, pp. 36-43.
5. Lin, C. E., Hsu, C. W., and Lee, Y. S. et al., "Verification of UAV Flight Control and Surveillance using Mobile Communication," AIAA Journal of Aerospace Computing, Information and Communication, Vol. 1, No. 4, April 2004, pp. 189-197.
6. Fernandes, P., and Nunes, U., "Vehicle Communications : A Short Survey," IADIS Telecommunications, Networks and Systems, Lisbon, 2007, pp. 134-138.
7. Hwang, D. H., Moon, S. W., Sung, T. K., Lee, S. J., "Design and implementation of an efficient tightly-coupled GPS/INS integration scheme", Proceedings of 2000 National Technical Meeting, 2000, pp. 159-165.
8. Zuberl, K. M., and Shin, K. G., "Scheduling Messages on Controller Area Network for Real-Time CIM Applications," IEEE Transactions on Robotics and Automation, Vol. 13, No. 2, pp.310-314, April 1997.
9. ISO 9141: Road Vehicles - Diagnostic System - Requirements for Interchange of Digital information, International Standard Organization, Oct. 1989.
10. Stojmenovic, I., "Geocasting with Guaranteed Delivery in Sensor Networks," IEEE Wireless Communications, Vol. 11, Issue:6, Dec. 2004, pp. 29-37.
11. Chen, C.T., Tekinay, S., and Papavassiliou, S., "Geocasting in Cellular Ad hoc Augmented Networks," Vehicular Technology Conference, Oct. 2003, Vol. 3, pp.1858-1862.

CONTACT INFORMATION

Chan Wei Hsu

winsonhsu@artc.org.tw

Tel:+886-4-7811222#2351

Fax:+886-4-7811333

Address: No.6, Lugong S. 7th Rd., Lugang Town, Changhua County 50544, Taiwan (R.O.C.)

http://www.artc.org.tw/index_en.aspx

ACKNOWLEDGMENTS

This work is supported in research projects 100-EC-17-A-04-02-0803 by Department of Industrial Technology, Ministry of Economic Affairs, Taiwan, R.O.C.

ABBREVIATIONS

GPS
global positioning system abbreviations

EMU
inertial measurement unit

INS
inertial navigation system

DSRC
dedicated short range communication

VU
vehicle unit

CAN
controller area network

UART
universal asynchronous receiver and transmitter

OBD
on-board diagnostic

CEP
circular error probable

RMS
root mean square

SAE International

Integrated Controller Design for Path Following in Autonomous Vehicles

2011-01-1032
Published
04/12/2011

Behrooz Mashadi , Pouyan Ahmadizadeh and Majid Majidi
Iran Univ. of Science and Tech.

doi:10.4271/2011-01-1032

ABSTRACT

This paper focuses on the problem of path following in autonomous vehicles. A solution to this problem is a strategy during which, a feedback/feed-forward control law is considered to control the path of the vehicle. In this controller, feedback and feed-forward gains are calculated based on the vehicle dynamic and road-concerned variables. A linear quadratic regulator (LQR) technique is applied to a three-degree of-freedom (DOF) linear vehicle model and three types of controllers are designed with the use of active front steering (AFS), four-wheel steering (4WS) and combination of an AFS with a direct yaw moment control (DYC) systems. A comparison is made about the performance and the efficacy among the different controllers in two different maneuvers. An 8-DOF nonlinear vehicle model is used for simulation purposes. It is concluded that 4WS system shows better potential for path following control of an autonomous vehicle.

KEYWORDS

path following, autonomous vehicles, optimal control, LQR, AFS, 4WS, DYC

INTRODUCTION

Automatic driving technology is the fundamental of intelligent transportation systems (ITS) which is studied by many researchers during recent years in order to reduce traffic problems. The primary task of automatic driving is to make an autonomous vehicle to follow a reference path automatically. During this path following, therefore, several issues should be considered to have an acceptable path control. Vehicle controllers that meet this requirement are called path followers. By applying appropriate controls which guide the vehicle along the path, the path following vehicle executes a globally defined geometric path. The goal of path following controller is to minimize the lateral distance between the vehicle and defined path, and to minimize the difference in the vehicle's heading and the defined path heading, and limit control inputs to smooth motions while maintaining the stability.

There are several studies which regard the path following problem. For instance, Snider [1] investigates path following by using optimal control theory. Front steering angle is taken as the control input to follow a desired path in low-g maneuvers. The path following control is treated by Goodarzi et al [2] by the application of linear quadratic regulator (LQR) technique. They use integrated active front steering (AFS) and direct yaw moment control (DYC) system. The AFS is used in low lateral accelerations and the DYC for high-g maneuvers wherein the tires are saturated and cannot produce enough lateral forces to control the vehicle on the path. This combination shows better performance during path following with more acceleration ranges need to be used.

Control theories other than linear techniques are also considered for the path following problem. For example Moriwaki [3] and You et al [4] both consider front steering system by application of robust control theory. El Hajjaji et al [5] use front steering system and fuzzy control theory to track a predetermined path.

This paper proposes a full optimal controller design process in which several controllers including AFS, four-wheel steering (4WS) and DYC are being used. The aim is to make a more realistic comparison for the performances among different individual control systems and their combinations, in order to obtain a useful and effective path following control.

The paper is organized as follows. Section 2 is devoted to vehicle dynamics modeling and formulation. Section 3 deals with design of controllers suitable for path following aims. The performance analysis of the controllers is considered in section 4 based on simulation analysis which is carried out by an 8-degree-of-freedom (DOF) nonlinear vehicle model generated in MATLAB environment. Finally, the work is concluded in section 5.

MODELING OF PATH FOLLOWING PROBLEM

In this section a dynamic model for the lateral vehicle motion is developed for controller design purposes [6], [7]. As shown in Figure 1, a 3-DOF vehicle model consisting of two displacement variables (x and y) and an angular orientation variable (ψ) is considered.

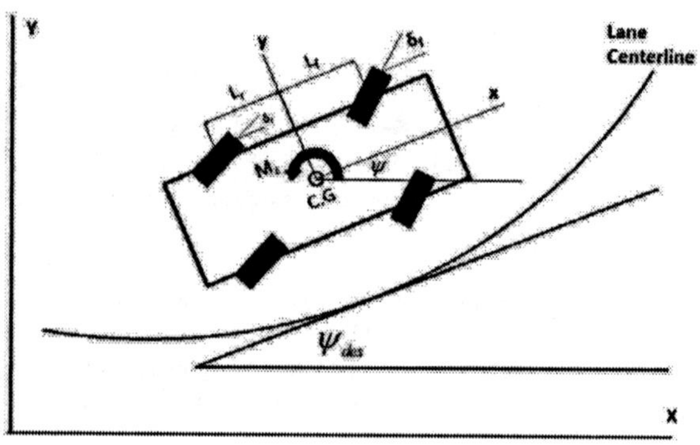

Figure1. Lateral Vehicle Dynamics

Considering a vehicle traveling with longitudinal speed u on a road of radius ρ, the rate of change of the desired orientation of the vehicle is defined by:

$$\dot{\psi}_{des} = \frac{u}{\rho} \tag{1}$$

The desired acceleration of the vehicle can be obtained from:

$$\frac{u^2}{\rho} = u\dot{\psi}_{des} \tag{2}$$

Two error parameters represented by the lateral distance of the C.G. of the vehicle from the centerline of the lane (e_1) and the orientation error of the vehicle with respect to the road (e_2) are considered. So $\ddot{e}_1$ can be obtained from:

$$\ddot{e}_1 = a_y - \frac{u^2}{\rho} = (\ddot{y} + u\dot{\psi}) - u\dot{\psi}_{des} = \ddot{y} + u(\dot{\psi} - \dot{\psi}_{des}) \tag{3}$$

where $\ddot{y}$ is the acceleration due to motion along the y-axis and $u\dot{\psi}$ is the centripetal acceleration.

By defining e_2 as:

$$e_2 = \psi - \psi_{des} \tag{4}$$

$\ddot{e}_1$ simply becomes:

$$\ddot{e}_1 = \ddot{y} + u\dot{e}_2 \tag{5}$$

In general for variable forward speed,

$$\dot{e}_1 = \dot{y} + \int u e_2 dt \tag{6}$$

This makes the model nonlinear and time varying which is not useful for control system design. So we assume that the longitudinal speed is constant in order to obtain a LTI model. For a constant u :

$$\dot{e}_1 = \dot{y} + u e_2 \tag{7}$$

Now, we consider the dynamic translational and rotational equations of a 2-DOF linear vehicle model:

$$m a_y = 2F_{yf} + 2F_{yr} \tag{8}$$

$$I_z \ddot{\psi} = (2L_f F_{yf} - 2L_r F_{yr}) + M_z \tag{9}$$

where M_z is the external yaw moment produced by DYC system and for small slip angles the lateral forces F_{yf} and F_{yr} are proportional to the slip angles. Hence, in order to model these forces, we define slip angles as:

$$\alpha_f = \delta_f - \theta_{vf} \tag{10}$$

$$\alpha_r = \delta_r - \theta_{vr} \tag{11}$$

where δ_f and δ_r are front and rear steering angles respectively. θ_{vf} and θ_{vr} are the angles that the velocity vectors at front and rear makes with the longitudinal axis of the vehicle and can be obtained by [8]:

$$\theta_{vf} = \frac{\dot{y} + L_f \dot{\psi}}{u} \tag{12}$$

$$\theta_{vr} = \frac{\dot{y} + L_r \dot{\psi}}{u}$$ (13)

where L_f and L_r are the distance of C.G. to front and rear axles respectively. The linear lateral tire forces for the front and rear wheels of the vehicle can therefore be written as:

$$F_{yf} = 2C_{\alpha f}(\delta_f - \theta_{vf})$$ (14)

$$F_{yr} = 2C_{\alpha r}(\delta_r - \theta_{vr})$$ (15)

Substituting from Eq. (14), (15), (4), (5) and (7) into (8) and (9), we find the state equations of vehicle motion as:

$$\dot{\mathbf{X}} = \mathbf{AX} + \mathbf{BU} + \mathbf{EW}$$ (16)

$$\mathbf{A} = \begin{bmatrix} 0 & 1 & 0 & 0 \\ 0 & a_{22} & a_{23} & a_{24} \\ 0 & 0 & 0 & 1 \\ 0 & a_{42} & a_{43} & a_{44} \end{bmatrix} \quad \mathbf{U} = \begin{Bmatrix} \delta_f \\ \delta_r \\ M_z \end{Bmatrix} \quad \mathbf{X} = \begin{Bmatrix} e_1 \\ \dot{e}_1 \\ e_2 \\ \dot{e}_2 \end{Bmatrix}$$

where

$$a_{22} = -2\frac{C_{\alpha f} + C_{\alpha r}}{mu} \qquad a_{23} = 2\frac{C_{\alpha f} + C_{\alpha r}}{m}$$

$$a_{24} = 2\frac{-C_{\alpha f}L_f + C_{\alpha r}L_r}{mu} \qquad a_{24} = 2\frac{-C_{\alpha f}L_f + C_{\alpha r}L_r}{I_z u}$$

$$a_{34} = 2\frac{C_{\alpha f}L_f - C_{\alpha r}L_r}{I_z} \qquad a_{44} = -2\frac{C_{\alpha f}L_f^2 + C_{\alpha r}L_r^2}{I_z u}$$

$$\mathbf{B} = \begin{bmatrix} 0 & 0 & 0 \\ \frac{2C_{\alpha f}}{m} & \frac{2C_{\alpha r}}{m} & 0 \\ 0 & 0 & 0 \\ \frac{2C_{\alpha f}L_f}{I_z} & -\frac{2C_{\alpha r}L_r}{I_z} & \frac{1}{I_z} \end{bmatrix}$$

$$\mathbf{E} = \begin{bmatrix} 0 \\ 2\frac{-C_{\alpha f}L_f + C_{\alpha r}L_r}{m} - u^2 \\ 0 \\ -2\frac{C_{\alpha f}L_f^2 + C_{\alpha r}L_r^2}{I_z} \end{bmatrix} \qquad \mathbf{W} = \left[\frac{1}{\rho}\right]$$

That is a general state-space equation with all the control inputs which has been derived for the path following problem. During controller designs in next session we use particular forms of this equation.

CONTROLLER DESIGN

The optimal control theory in the form of LQR will be applied to the problem in order to meet the path following objectives in this paper [9], [10]. Different systems like AFS, 4WS and DYC will be examined in order to investigate their influence on the vehicle path following property. Obviously, the desired performance of the controllers is to keep the errors e_1 and e_2 minimal so that the vehicle can best follow the path. The performance is assessed by a scalar performance index, of the form:

$$J = \int_0^{\infty} [(\mathbf{X} - \mathbf{X_d})^T \mathbf{Q} (\mathbf{X} - \mathbf{X_d}) + \mathbf{U}^T \mathbf{RU}]\,dt$$ (17)

which will be called index of path following. In this index **X** and **U** are the state and the input vectors respectively. $\mathbf{X_d}$ is the desired values of states which all are considered zero because we want the vehicle follow exactly the path without any heading angle. So, it requires that e_{1d} and e_{2d} to be zero as well as their derivatives. **Q** and **R**, both symmetric matrices, are positive semi definite and positive definite respectively. **Q** is a diagonal weighting matrix with an entry for each state corresponding to the performance aspects contributing to the cost function and **R** is weighting value corresponding to control effort contributing to the cost function.

Considering the LTI system as:

$$\dot{\mathbf{X}} = \mathbf{AX} + \mathbf{BU} + \mathbf{EW}$$ (18)

with initial state, and the control law:

$$\mathbf{U} = -\mathbf{K_{fb}X} - \mathbf{K_{ff}W}$$ (19)

The state feedback gain matrix $\mathbf{K_{fb}}$ and the disturbance feed-forward gain matrix $\mathbf{K_{ff}}$ can be determined such that the performance index is minimal. Stabilizing feedback gain $\mathbf{K_{fb}}$ for minimizing the performance index can be calculated by:

$$\mathbf{K_{fb}} = \mathbf{R}^{-1}\mathbf{B}^T\mathbf{P}$$ (20)

where **P** satisfies the matrix Riccati equation:

$$\mathbf{A}^T\mathbf{P} + \mathbf{PA} - \mathbf{PBR}^{-1}\mathbf{B}^T\mathbf{P} + \mathbf{Q} = \mathbf{0}$$ (21)

and the feed-forward gain matrix $\mathbf{K_{ff}}$ minimizing the performance index can be calculated by:

$$\mathbf{K}_{ff} = \mathbf{R}^{-1}\mathbf{B}^T\mathbf{S} \tag{22}$$

where **S** satisfies the following matrix equation:

$$(\mathbf{A}^T - \mathbf{PBR}^{-1}\mathbf{B}^T)\mathbf{S} + \mathbf{KE} = \mathbf{0} \tag{23}$$

again **P** is calculated from the Ricatti equation [11], [12].

Taking into consideration the matrices **A**, **B**, and **E** one can see that the feedback gain matrix $\mathbf{K_{fb}}$ is a matrix with 4 columns, because of having 4 states, and the number of rows depends on the number of control inputs. Also the number of control inputs determines the number of feed-forward gain matrix $\mathbf{K_{ff}}$.

AFS CONTROLLER

For this controller only front steering angle is available as control input. So, the weighting matrices **Q**, **R**, and the matrix **B** take the following forms:

$$\mathbf{Q} = \begin{bmatrix} W_1 & 0 & 0 & 0 \\ 0 & W_2 & 0 & 0 \\ 0 & 0 & W_3 & 0 \\ 0 & 0 & 0 & W_4 \end{bmatrix} \qquad \mathbf{R} = [W_5]$$

$$\mathbf{U} = \{\delta_f\} \qquad \mathbf{B} = \begin{bmatrix} 0 \\ 2\frac{C_{\alpha f}}{m} \\ 0 \\ 2\frac{C_{\alpha f} L_f}{I_z} \end{bmatrix}$$

With the use of weighting matrices **Q** and **R** using the Eq. (20) and Eq. (22), the feedback and feed-forward gain matrices can be calculated; consequently, the measure of control input can be achieved by:

$$\delta_f = K_{fb_1} e_1 + K_{fb_2} \dot{e}_1 + K_{fb_3} e_2 + K_{fb_4} \dot{e}_2 + K_{ff_1}(\frac{1}{\rho}) \tag{24}$$

The determination of weighting factors of matrices **Q** and **R** is a trial and error process and this tuning should be carried out to ensure good responses at the output.

4WS CONTROLLER

For this controller, both the front and the rear steering angles are considered as control inputs. Thus the result is:

$$\mathbf{Q} = \begin{bmatrix} W_1 & 0 & 0 & 0 \\ 0 & W_2 & 0 & 0 \\ 0 & 0 & W_3 & 0 \\ 0 & 0 & 0 & W_4 \end{bmatrix} \qquad \mathbf{R} = \begin{bmatrix} W_5 & 0 \\ 0 & W_6 \end{bmatrix}$$

$$\mathbf{U} = \begin{Bmatrix} \delta_f \\ \delta_r \end{Bmatrix} \qquad \mathbf{B} = \begin{bmatrix} 0 & 0 \\ 2\frac{C_{\alpha f}}{m} & 2\frac{C_{\alpha r}}{m} \\ 0 & 0 \\ 2\frac{C_{\alpha f} L_f}{I_z} & -2\frac{C_{\alpha r} L_r}{I_z} \end{bmatrix}$$

So the control inputs are:

$$\delta_f = K_{fb_1} e_1 + K_{fb_2} \dot{e}_1 + K_{fb_3} e_2 + K_{fb_4} \dot{e}_2 + K_{ff_1}(\frac{1}{\rho}) \tag{25}$$

$$\delta_r = K'_{fb_1} e_1 + K'_{fb_2} \dot{e}_1 + K'_{fb_3} e_2 + K'_{fb_4} \dot{e}_2 + K_{ff_2}(\frac{1}{\rho}) \tag{26}$$

AFS+DYC CONTROLLER

For this case the front steering angle and the yaw moment are available as the control inputs. Therefore:

$$\mathbf{Q} = \begin{bmatrix} W_1 & 0 & 0 & 0 \\ 0 & W_2 & 0 & 0 \\ 0 & 0 & W_3 & 0 \\ 0 & 0 & 0 & W_4 \end{bmatrix} \qquad \mathbf{R} = \begin{bmatrix} W_5 & 0 \\ 0 & W_7 \end{bmatrix}$$

$$\mathbf{U} = \begin{Bmatrix} \delta_f \\ M_z \end{Bmatrix} \qquad \mathbf{B} = \begin{bmatrix} 0 & 0 \\ 2\frac{C_{\alpha f}}{M} & 0 \\ 0 & 0 \\ 2\frac{C_{\alpha f} L_f}{I_z} & \frac{1}{I_z} \end{bmatrix}$$

And the control inputs are:

$$\delta_f = K_{fb_1} e_1 + K_{fb_2} \dot{e}_1 + K_{fb_3} e_2 + K_{fb_4} \dot{e}_2 + K_{ff_1}(\frac{1}{\rho}) \tag{27}$$

$$M_z = K''_{fb_1} e_1 + K''_{fb_2} \dot{e}_1 + K''_{fb_3} e_2 + K''_{fb_4} \dot{e}_2 + K_{ff_3}(\frac{1}{\rho}) \tag{28}$$

SIMULATION AND PERFORMANCE ANALYSIS

In this section, in order to evaluate the performance of the designed controller, simulations using an 8-DOF nonlinear vehicle model constructed in MATLAB software are used. For the simulations, two desired paths are considered:

- A mid-lateral acceleration maneuver during which the vehicle experiences a J-turn maneuver with a constant forward speed of 36 m/s.
- A high-lateral acceleration maneuver during which the vehicle experiences a Lane-Change maneuver with constant longitudinal speed of 45 m/s.

The chosen vehicle is a typical passenger car with information given in Table 1. The simulation results for the two above described maneuvers are presented in Figures 2 and 3. Three types of controllers are examined, the AFS controller that uses only front steering angle, the 4WS controller with both front and rear steering angles and the AFS+DYC controller that uses front steering angle together with correcting yaw moment.

The variations of vehicle lateral acceleration yaw rate and sideslip angle for the first maneuver are shown in Figure 2a, 2b, 2c, 2d, 2e, 2f, 2g. The overall performance of the vehicle for all three controllers is close. Figure 2d shows the lateral distance of the vehicle C.G. from the lane centerline, or the lateral offset (e_1). For 4WS controller the values of this parameter are seen to be much lower compared to the results of the two other systems. The orientation error (e_2) depicted in Figure 2e shows almost similar variations for all controllers with small differences. Figure 2f illustrates the front steering angles for the three systems. The peak values are all in the same order but 4WS uses larger values and AFS +DYC smallest values throughout the maneuver. Figure 2g belongs only to the rear steering angle of the 4WS system whereas the variation of external yaw moment applied to the vehicle in the AFS+DYC system is shown in Figure 2h.

For the second maneuver the results are plotted in Figure 3a, 3b, 3c, 3d, 3e, 3f, 3g, 3h, 3i. Figures 3a, 3b, 3c, 3d show the variations of the vehicle path, lateral acceleration, yaw rate and sideslip angle respectively. The AFS+DYC path seems to be closer to the desired path, however, the two errors e_1 and e_2 shown in Figures 3e and 3f reveal that the 4WS system have better overall errors due mainly to very small orientation errors. Front steering angles for the two cases are almost similar (see Figure 3g) but rear steering angle for 4WS shown in Figure 3h play the important role. Despite the large yaw moments applied by the AFS+DYC system, seen in Fig. 3i, the 4WS system is more successful since the application of front and rear steers results in producing a proper yaw moment together with sufficient lateral force which helps the vehicle to better follow the path. Because of the path nature during the lane change maneuver in high lateral accelerations, the differences of the performances for the three systems were seen better in this maneuver.

From the foregoing discussion it can be concluded that the application of 4WS system for path following problem in autonomous vehicles shows better results in comparison to AFS and AFS+DYC systems. In other words, in path following problem, the steerability of all wheels provides enough yaw moment as well as proper lateral forces which keeps the vehicle on the desired path. At DYC systems the external yaw moment is produced by the application of differential braking which has undesired effects on the longitudinal dynamics of the vehicle.

Table 1. Vehicle data used in simulations

Vehicle wheelbase	$L = 2.5\ (m)$
Distance from C.G. to front axle	$L_f = 1.2\ (m)$
Distance from C.G. to rear axle	$L_r = 1.3\ (m)$
Front tire stiffness	$C_{\alpha f} = 35000\ (N/rad)$
Rear tire stiffness	$C_{\alpha r} = 35000\ (N/rad)$
Yaw moment of inertia	$I_z = 2500\ (Kg.m2)$
Vehicle mass	$m = 1300\ (Kg)$
Coefficient of tire-road friction	$\mu = 0.9$
Road radius of curvature	$\rho = 300\ (m)$

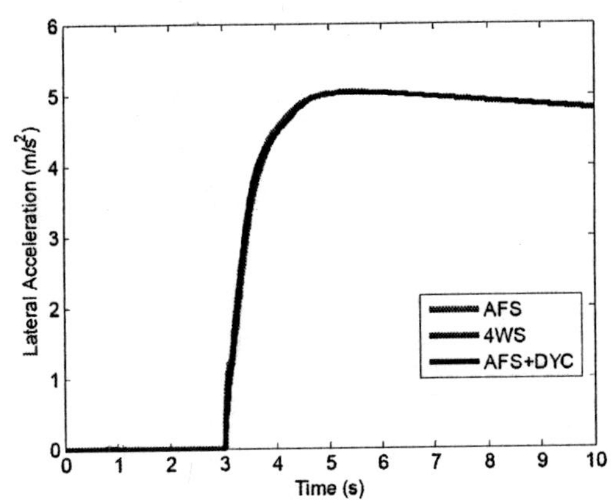

(a)

Fig2. Simulation results for J-turn maneuver: a) Lateral Acceleration, b) Yaw Rate, c) Slip Angle, d) Lateral distance error, e) Orientation error, f) Front Steering Angle, g) Rear Steering Angle, h) Yaw Moment

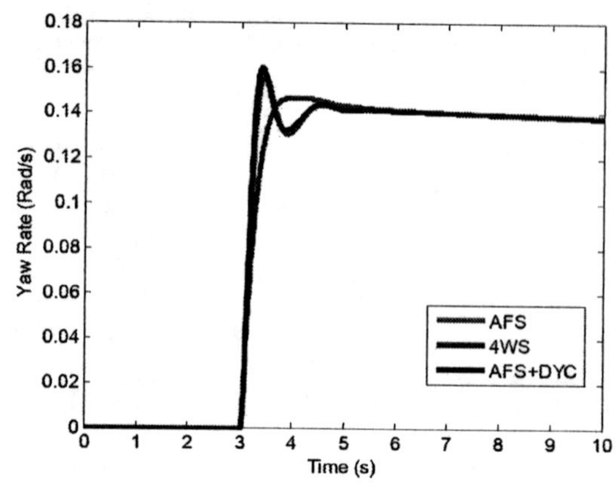

(b)

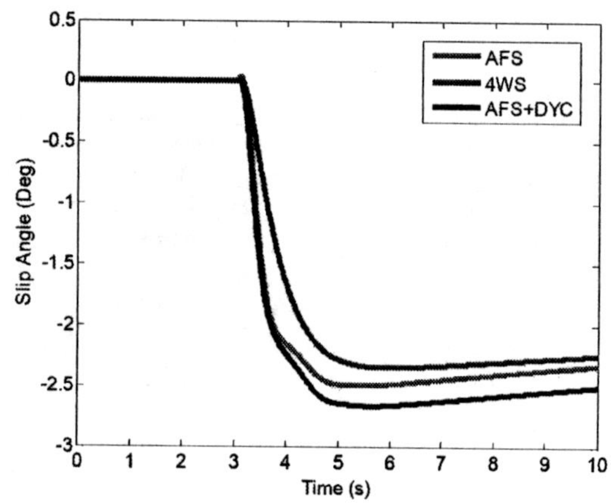

(c)

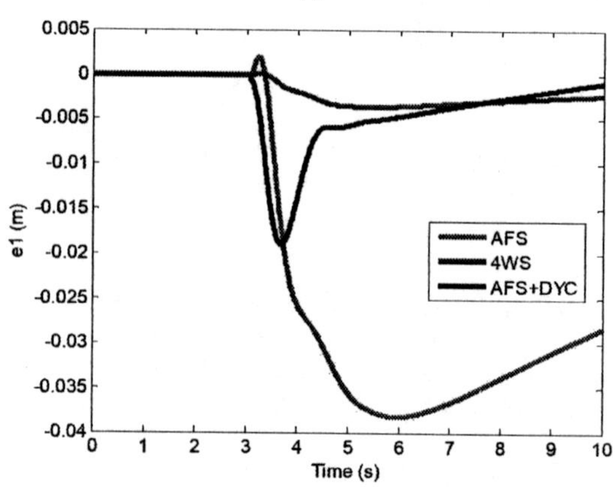

(d)

Fig2 (cont.). Simulation results for J-turn maneuver: a) Lateral Acceleration, b) Yaw Rate, c) Slip Angle, d) Lateral distance error, e) Orientation error, f) Front Steering Angle, g) Rear Steering Angle, h) Yaw Moment

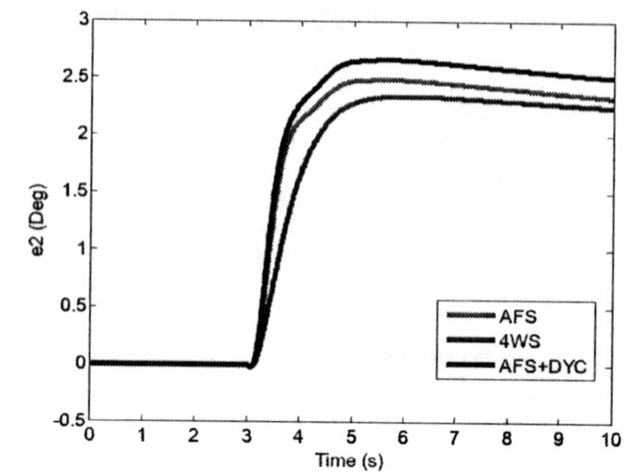

(e)

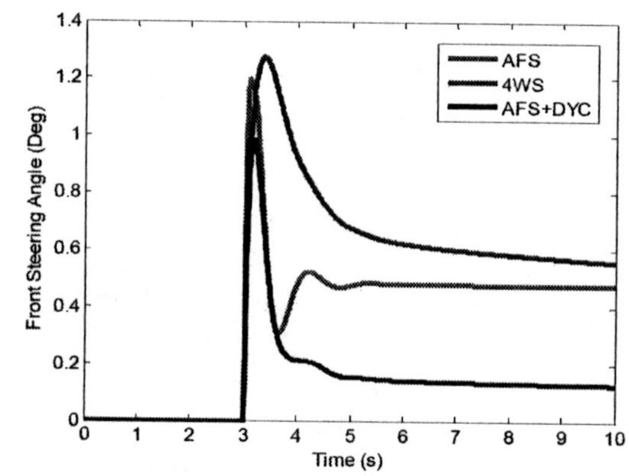

(f)

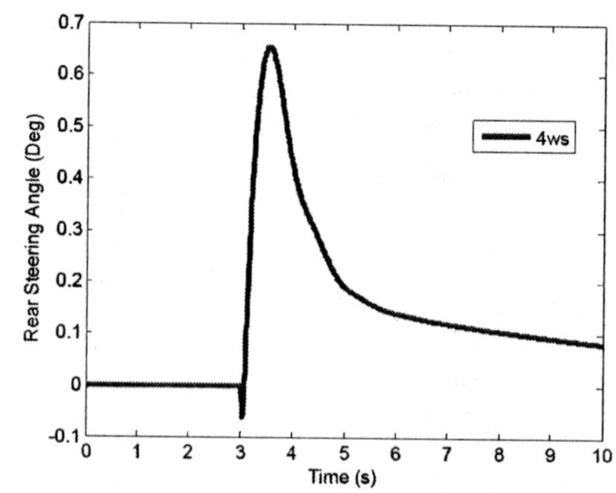

(g)

Fig2 (cont.). Simulation results for J-turn maneuver: a) Lateral Acceleration, b) Yaw Rate, c) Slip Angle, d) Lateral distance error, e) Orientation error, f) Front Steering Angle, g) Rear Steering Angle, h) Yaw Moment

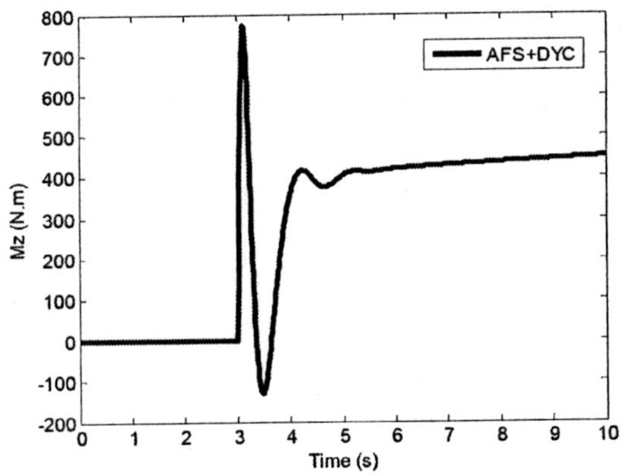

(h)

Fig2 (cont.). Simulation results for J-turn maneuver: a) Lateral Acceleration, b) Yaw Rate, c) Slip Angle, d) Lateral distance error, e) Orientation error, f) Front Steering Angle, g) Rear Steering Angle, h) Yaw Moment

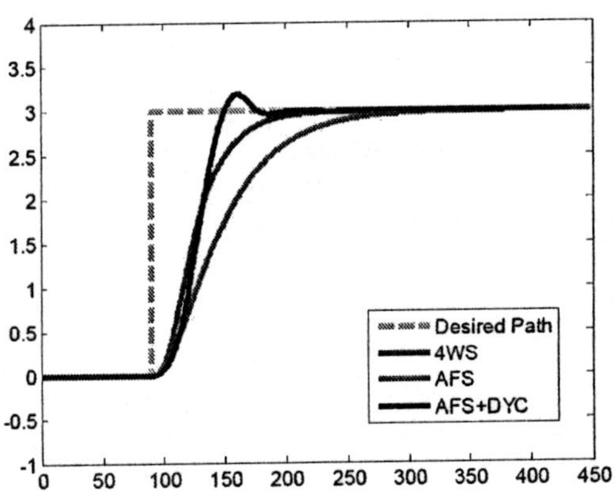

(a)

Figure 3. Simulation results for Lane-Change maneuver: a) Vehicle path, b) Lateral Acceleration, c) Yaw Rate, d) Slip Angle, e) Lateral distance error, f) Orientation error, g) Front Steering Angle, h) Rear Steering Angle, i) Yaw Moment

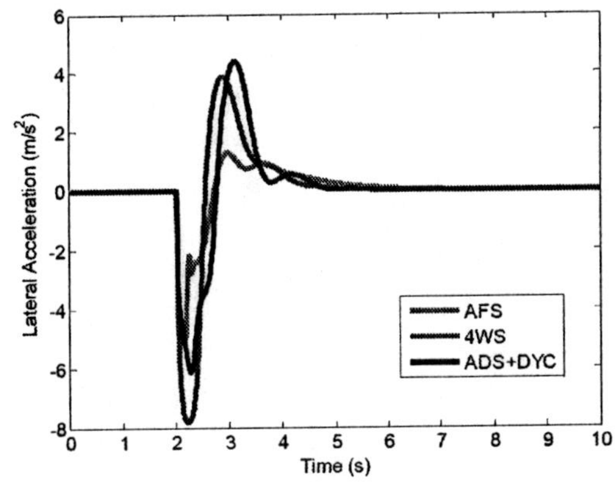

(b)

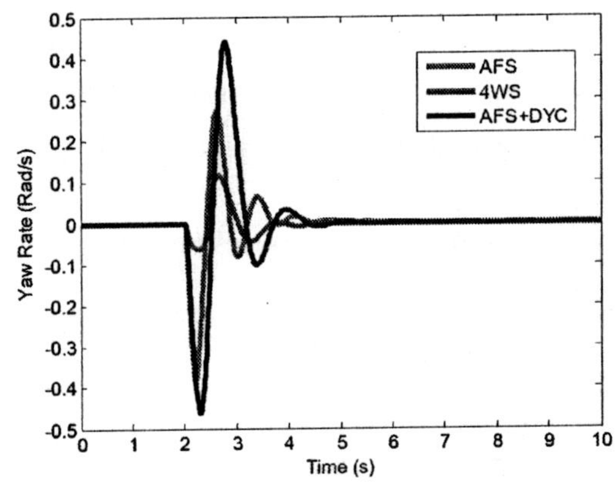

(c)

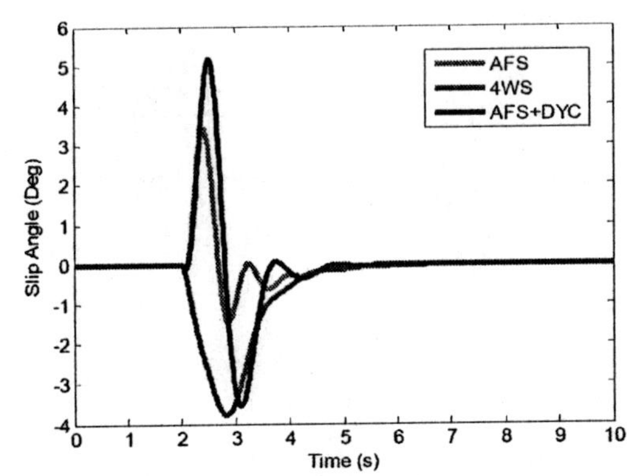

(d)

Figure 3 (cont.). Simulation results for Lane-Change maneuver: a) Vehicle path, b) Lateral Acceleration, c) Yaw Rate, d) Slip Angle, e) Lateral distance error, f) Orientation error, g) Front Steering Angle, h) Rear Steering Angle, i) Yaw Moment

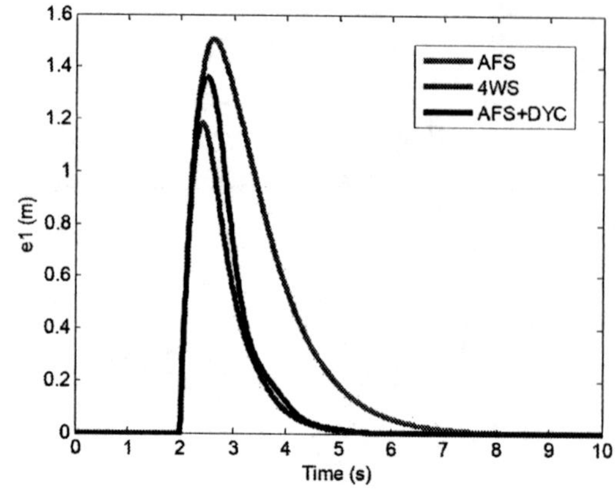

(e)

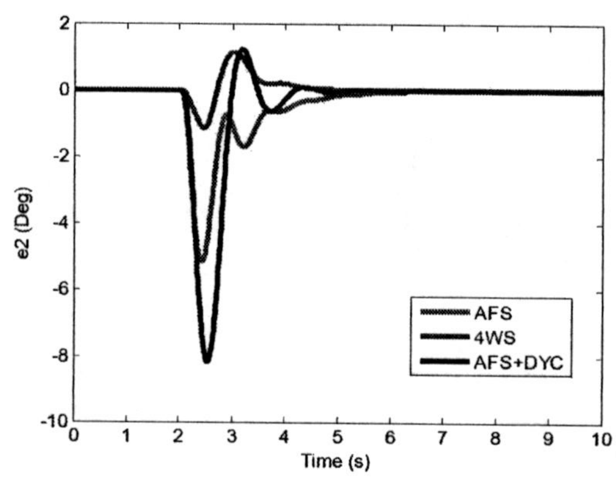

(f)

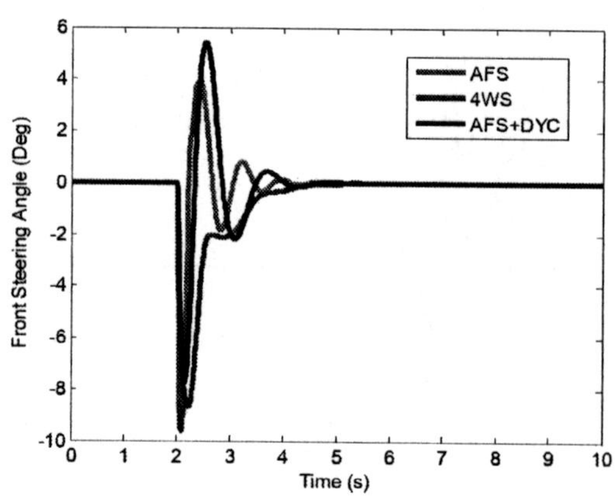

(g)

Figure 3 (cont.). Simulation results for Lane-Change maneuver: a) Vehicle path, b) Lateral Acceleration, c) Yaw Rate, d) Slip Angle, e) Lateral distance error, f) Orientation error, g) Front Steering Angle, h) Rear Steering Angle, i) Yaw Moment

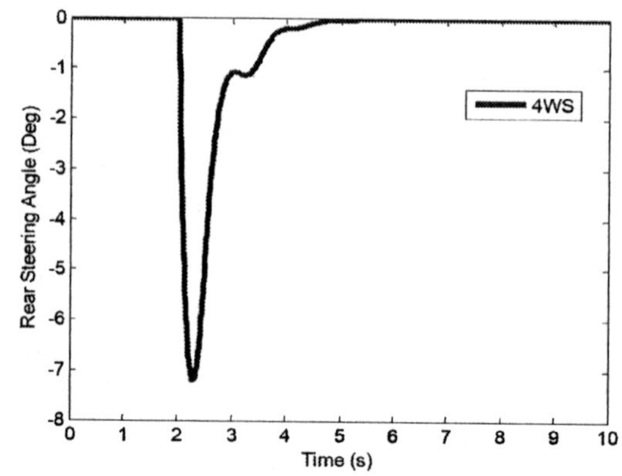

(h)

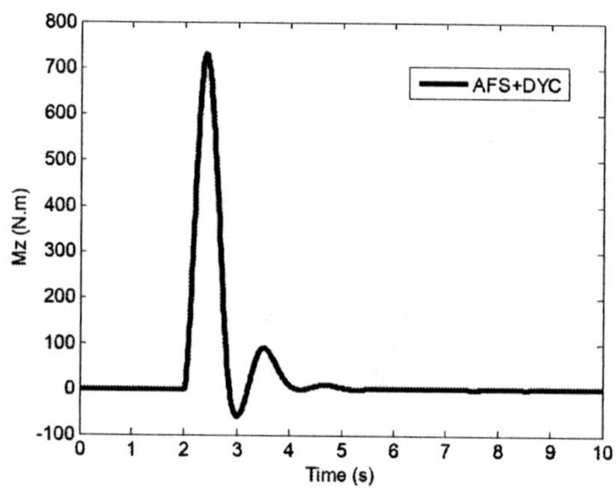

(i)

Figure 3 (cont.). Simulation results for Lane-Change maneuver: a) Vehicle path, b) Lateral Acceleration, c) Yaw Rate, d) Slip Angle, e) Lateral distance error, f) Orientation error, g) Front Steering Angle, h) Rear Steering Angle, i) Yaw Moment

CONCLUSION

An optimal control strategy has been proposed in this paper in order to control an autonomous vehicle on a specified path. Based on this strategy, three controllers were designed using AFS only, 4WS only and AFS+DYC systems. The comparisons among these controllers were made from the results of numerical simulations by using an 8-DOF nonlinear vehicle model. The results showed that the performance of the 4WS system was better than the two other systems in terms of having lower lateral offset and orientation errors. The AFS+DYC system also had responses close to that of 4WS system especially for J-turn maneuver; however, it required large yaw moments resulting from differential braking which can have undesired effects on the longitudinal dynamics of the vehicle.

REFERENCES

1. Snider, J.M., "Automatic Steering Methods For Autonomous Automobile Path Tracking," Ph.D. thesis, Robotics Institute, Carnegie Mellon University, Pittsburg, Pennsylvania, 2009.

2. Goodarzi, A., Sabooteh, A., Esamailzadeh, A., "Automatic Path Control Based on Integrated Steering and External Yaw-moment Control," *Proc. of ImechE, Part K: J. of Multi-body dynamics* **222**(2), 2008, doi:10.1243/14644193JMBD120.

3. Moriwaki, K., "Autonomous Steering Control for Electric Vehicles Using Nonlinear State Feedback H∞ Control," *Nonlinear Analysis* **63**(5-7): 2257-2268, 2005, doi:10.1016/j.na.2005.03.065.

4. You, S. S., Jeon, S. K., "Controller Design and Analysis for Automatic Steering of Passenger Cars," *Mechatronics* **12**(3): 427-446, 2002.

5. El Hajjaji, A., Bentalba, S., "Fuzzy Path Tracking Control for Automatic Steering of Vehicles," *Robotics and Autonomous Systems* **43**(4):203-213, 2003, doi:10.101/S0921-8890(03)00003-4.

6. Consolini, L.; Piazzi, A., Tosques, M., "Path Following of Car-like Vehicles Using Dynamic Inversion," *International Journal of Control* **76**(17): 1724-1738, 2003, doi: 10.1080/00207170310001633303.

7. Balluchi, A., Bicchi, A., Souères, P., "Path-following with Abounded-curvature Vehicle: a hybrid control approach," *International Journal of Control* **78**(15): 1228-1247, 2005, doi:10.1080/00207170500275096.

8. Yu, S. H., Moskwa, J. J., "A Global Approach to Vehicle Control: Coordination of Four Wheel Steering and Wheel Torques," *Journal of Dynamic Systems, Measurement, and Control* **116**(4):659-667, 1994.

9. Lim, E. M., "Lateral and Longitudinal Vehicle Control Coupling in the Automated Highway System," M.SC. Thesis, Mechanical Engineering Department, University of California at Berkeley, California, 1998.

10. Smith, D., Benton, R., "Nonlinear-gain-optimized Controller Development and Evaluation for Automated Emergency Vehicle Steering," *International Journal of Vehicle Design* **24**(1):79-99, 2000.

11. Belanger, P., "Control Engineering, a Modern Approach," Saunders College Publication, Philadelphia, ISBN 0-03-013489-7: 291-358, 1995.

12. MATLAB (Version 7.9), Documentation and Help, Computer Software, Mathworks, 2009.

13. Lauffenburger, J. PH., Basset, M., Coffin, F., and Gissinger, G. L., "Driver-aid System Using Path-planning for Lateral Vehicle Control," *Control Engineering Practice* **11**(2):217-231, 2003.

NOTATION

C.G.
Center of gravity

m
Vehicle mass

L_f / L_r
C.G. distance to Front/Rear axle

C_{af} / C_{ar}
Front/Rear tire cornering stiffness

I_z
Moment of inertia about vehicle z axis

u
Longitudinal velocity

F_{yf} / F_{yr}
Front/Rear lateral tire force

ψ
Yaw angle

a_y
Lateral acceleration

ρ
Road radius

α_f / α_r
Front/Rear tire sideslip angle

δ_f / δ_r
Front/Rear steering angle

$\theta_{vf} / \theta_{vr}$
Velocity angle with longitudinal axis at front/rear

M_z
External moment of inertia

$W_1 .. W_7$
Weighting factors

2011-28-0035

Autonomous Vehicle Control in Urban Environment by Map-Based Driving Lane Detection

Takanori Yoshizawa, Pongsathorn Raksincharoensak, Masao Nagai
Tokyo University of Agriculture and Technology

ABSTRACT

Highly-precise ego-localization and mapping techniques from the road shape features are key elements in order to realize an autonomous driving system for vehicle in urban area which has complex environments. The objective of this study is to develop an autonomous driving system based on mapping and ego-localization using a LIDAR. To handle curved path tracking scenario, this paper proposes a desired steering angle generator considering a constructed map using the LIDAR in real time combined with the feedback control of the preview lateral deviation. The effectiveness of the proposed control method is verified by simulation and test drives using the autonomous path tracking control system.

INTRODUCTION

Statistical data reveal that single accidents of vehicles with especially high fatality rate occur due to the driver's steering wheel misoperation [1]. Autonomous driving systems have been developed in order to improve traffic safety and drive comfort in recent years. The autonomous driving system is expected to be one of promising traffic accident prevention approaches as it can reject human error such as error of judgment and lack of environment information perception that are principal cause factors of traffic accidents. Computer-controlled vehicles are one of promising approaches for improving road capacity, as well as reducing traffic congestion and air pollution. Furthermore, as one of social demands for future mobility, low-speed autonomous driving systems in urban area are required in order to support mobility and quality of life (QOL) of elderly people and assist the driving of elderly drivers in high-risk critical driving situations.

In urban area, knowledge about the road shape and precise ego-localization are important key elements for autonomous driving systems. The accuracy in order to realize the autonomous driving system is required to be a few-centimeter order. However, the positional accuracy of stand-alone GPS technology for car-navigation systems is at 20-30-meter order. Hence, the accuracy of stand-alone GPS system is insufficient for the autonomous driving system. Therefore, a RTK-GPS (Real Time Kinematic-GPS) is one of the methods in order to estimate precise ego-localization, and the autonomous driving system using RTK-GPS as well as the data fusion with other sensors have been developed [2-3]. However, the accuracy of RTK-GPS depends on the infrastructure that transmits correction signal to correct the vehicle position. In addition, due to the expensive cost of RTK-GPS, it is still difficult in current status to equip the RTK-GPS in all vehicles for economical reasons.

From the abovementioned background, the authors have proposed the autonomous driving system using a LIDAR which measures the precise relative distance from ego-vehicle to surrounding objects [4]. This system calculate desired steering angle for path tracking based on lateral position of right and left curbs at the preview point detected by using a LIDAR. As the curb position is detected at the preview point, the detected curb becomes discontinuous when the vehicle approaches a sharp curve. As a result, the previously proposed system has a problem that there is oscillation of the desired yaw rate during cornering. Therefore, In order to handle this situation, this paper adds a trapezoidal steering angle term to the pure feedback control system in order to secure the stability of the vehicle motion during cornering.

The rest of the paper is organized as follows: Section 2 describes the configuration of the experimental vehicle. Next, Section 3 describes the vehicle model used in the path tracking simulation and experiments. Section 4 and 5 describes the design of the autonomous path tracking control system based on LIDAR information. Section 6 describes the newly proposed method to generate the desired steering angle input which is trapezoid shape. Section 7 and 8 describe the simulation and test drives of the newly proposed autonomous vehicle control system in order to make a right turn at a junction. Finally, the major conclusions to be drawn from the study are summarized in Section 9.

EXPERIMENTAL VEHICLE

The micro electric vehicle used in this research is shown in Fig. 1. The LIDAR is equipped at the front section of the vehicle, tilted downward to the ground, in order to detect road boundaries in a detection range of

180 degrees. Additionally, the sensors which measure the velocity, the longitudinal and lateral accelerations and the yaw rate are also mounted. A digital signal processing system which is equipped on the vehicle is used to control driving torques of in-wheel motors and steering wheel angle with AC-servo motor in real time. In addition, the in-wheel motors generate driving torque depending on the command voltage signal and the AC-servo motor generates steering torque to control the steering wheel angle by the pulse width control with a servo amplifier.

VEHICLE MODEL

LONGITUDINAL VEHICLE DYNAMICS - The longitudinal vehicle model used in this research is shown in Fig. 2. The governing equation of longitudinal motion can be expressed as follows:

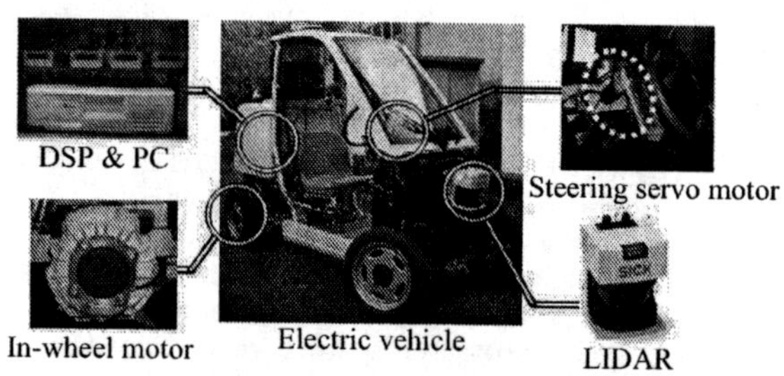

Fig. 1 Configuration of sensors and actuators of experimental vehicle

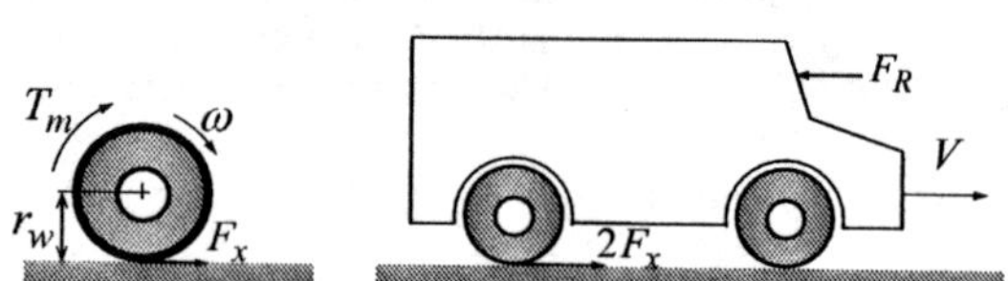

(a) 1-wheel rotational model (b) Longitudinal vehicle model

Fig. 2 Longitudinal vehicle dynamics

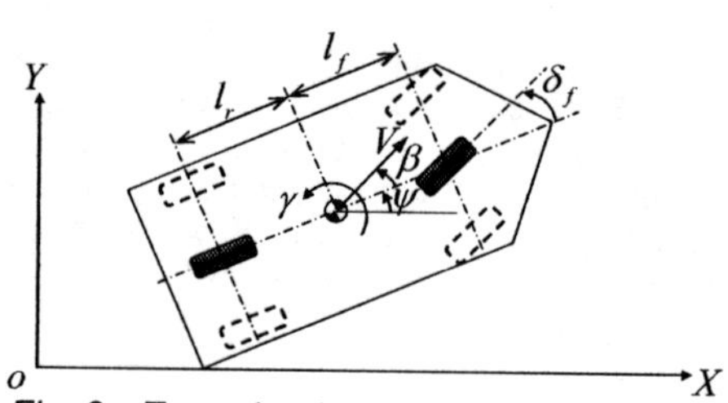

Fig. 3 Two-wheel model in lateral motion

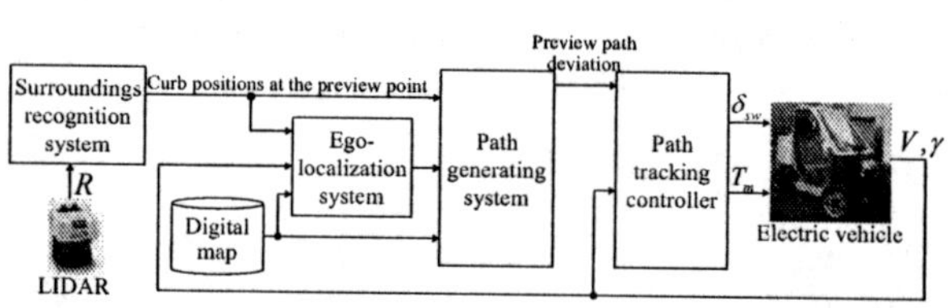

Fig. 4 Schematic diagram of autonomous path tracking control system

$$J\dot{\omega} = T_m - F_x r_w \tag{1}$$

$$\omega = \frac{V}{r_w} \tag{2}$$

$$m\dot{V} = 2F_x - F_R \tag{3}$$

where, J is the inertia moment of wheel, ω is wheel velocity, T_m is driving torque, F_x is longitudinal force, r_w is the effective tire radius, V is vehicle velocity, m is the vehicle mass and F_R is the rolling resistance.

LATERAL VEHICLE DYNAMICS - The two-wheel vehicle model in the lateral motion is shown in Fig. 3.The linear equivalent two-wheel vehicle model in lateral motion can be expressed as follows:

$$\begin{bmatrix}\dot{\beta}\\ \dot{\gamma}\end{bmatrix} = \begin{bmatrix}a_{11} & a_{12}\\ a_{21} & a_{22}\end{bmatrix}\begin{bmatrix}\beta\\ \gamma\end{bmatrix} + \begin{bmatrix}b_{11}\\ b_{21}\end{bmatrix}\frac{\delta_{sw}}{N} \tag{4}$$

$$a_{11} = -\frac{2(C_f + C_r)}{mV} \qquad a_{12} = -\left(1 + \frac{2(C_f l_f - C_r l_r)}{mV^2}\right)$$

$$a_{21} = -\frac{2(C_f l_f - C_r l_r)}{I_z} \qquad a_{22} = -\frac{2(C_f l_f^2 + C_r l_r^2)}{I_z V}$$

$$b_{11} = \frac{2C_f}{mV} \qquad b_{21} = \frac{2C_f l_f}{I_z}$$

where, C_f, C_r indicate the front and rear cornering stiffness respectively, l_f, l_r are distances between the vehicle center and the front and rear axes respectively, and I_z is the yaw inertia moment of vehicle.

The current position of the center of gravity of the vehicle in the earth-fixed coordinate system (X, Y) and the current yaw angle of the vehicle ψ are shown as the following mathematical representations.

$$\begin{cases} X(t) = X_0 + \int_0^t V(t)\cos(\psi(t) + \beta(t))dt \\ Y(t) = Y_0 + \int_0^t V(t)\sin(\psi(t) + \beta(t))dt \\ \psi(t) = \psi_0 + \int_0^t \gamma(t)dt \end{cases} \tag{5}$$

AUTONOMOUS PATH TRACKING CONTROL SYSTEM

The overview of the autonomous path tracking control system proposed at the previous paper is shown as Fig. 4. The autonomous vehicle is driven based on the assumption that a map-based road boundary is acquired beforehand, e.g. by using LIDAR to detect road curbs, walls, or lane markers on both sides of the vehicle. In addition, the detected road boundary information is prefiltered by geometric approximation heuristically to reject measurement noise and data loss during detection process which might result in unstable vehicle control. The relative distance data between obstacles and the experimental vehicle is acquired by using LIDAR when the experimental vehicle was driven by a human driver. A digital map is constructed based on landmarks such as both side curbs of road, poles, road signs and walls of buildings which are detected by the distance data acquired by using LIDAR. The constructed digital map and surroundings environment acquired by using LIDAR are used for ego-localization. Then the target path is determined by the digital map. To follow the target path, the in-wheel motor and the steering wheel motor are controlled based on the desired yaw rate which is calculated from the desired lateral displacement at the preview point. As the curb position is detected at the preview point, the detected curb becomes discontinuous when the vehicle approaches a sharp corner as shown in Fig.5. Then, the previously proposed system has a problem that there is unsatisfactory oscillation of the desired yaw rate during cornering.

Therefore, in this paper, the desired steering angle generating system is proposed to improve the curved path tracking performance during right-angle cornering. In the straight pathway, the previously path tracking control system based on the feedback of road boundary position, i.e. curb position, is used to track the straight pathway. In the right-angle cornering maneuver with discontinuous curb detection by using a LIDAR, a feedforward control system which determines the steering angle profile for cornering is used.

Fig. 5 Road used for LIDAR data acquisition

PATH TRACKING CONTROL SYSTEM DESIGN

FEEDBACK CONTROL SYSTEM DESIGN - Fig. 6 shows the coordinates of the road boundary in the vehicle-fixed coordinate system. The control target of the path tracking is to regulate the desired lateral deviation at the determined preview point. In this report, the lateral deviation from the road center at the preview point is calculated based on the relative distance of the road curbs and the vehicle front end detected by using the LIDAR. The desired yaw rate which is required to regulate the lateral deviation is theoretically calculated from the lateral deviation. Then, the desired yaw rate is used as the command signal to control the steering wheel via the steering actuator.

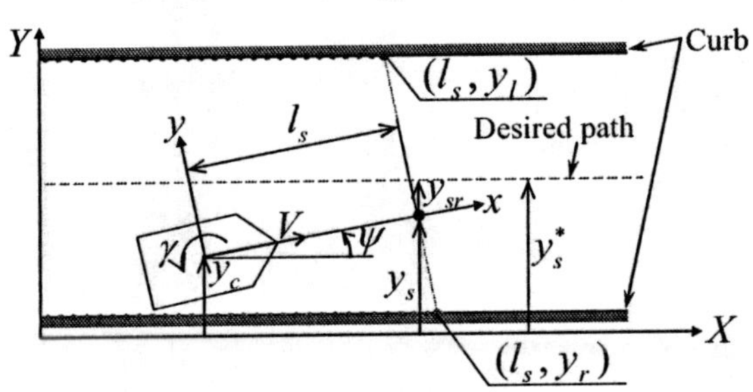

Fig. 6 Path tracking control by the previous method

PREVIEW PATH DEVIATION CALCULATION - The preview path deviation, y_{sr}, that is the lateral deviation from the road center to the preview point is shown as the following equation.

$$y_{sr}(t) = \frac{y_r(t) + y_l(t)}{2} \tag{6}$$

where, t indicates the time, y_r, y_l are the positions of right and left curbs at the vehicle-fixed coordinate system. In this paper, y_{sr} is the average of lateral position of the right and left curbs at the vehicle-fixed coordinate system.

DESIRED YAW RATE CALCULATION – As the objective of the path tracking control, the desired lateral displacement at the preview point and the predicted lateral displacement must be equal. Thus, as the following equation (7), the desired yaw rate "γ^*" can be calculated by the preview path deviation "y_{sr}" as shown in the previous report [4]. y_{sr} is the lateral deviation from the desired path "y_s^*" to the preview point "y_s" as shown in Fig. 6.

$$\gamma^*(t) = \frac{2V}{l_s^2} y_{sr}(t) \tag{7}$$

where, l_s is the preview distance. l_s is the calculated by the vehicle velocity "V" and the predicted time "T" as follows:

$$l_s = VT \tag{8}$$

DESIRED STEERING WHEEL ANGLE CALCULATION - We assume that the frequency region of steering wheel input for the autonomous driving is low compared to the emergency evasive steering, so the dynamic characteristics can be neglected and does not influence the controller performance. By using the relationship between the steady-state yaw rate and the steering angle described in the linear equivalent two-wheel vehicle model, the steering wheel angle can be calculated from the desired yaw rate as follows:

$$\delta_{sw} = n(1 + KV^2)\frac{l}{V}\gamma^* \tag{9}$$

where, n is the steering gear ratio, K is the stability factor of the vehicle and l is the wheelbase. Here, the steering wheel angle is controlled by the servo amplifier using its position control function.

FEEDFORWARD CONTROL SYSTEM BASED ON PRIORI CORNER INFORMATION

SYSTEM OVERVIEW - The proposed autonomous driving system obtains the curve position by a few meter orders on condition that the proposed system uses stand-alone GPS system for car-navigation systems. However, the accuracy of stand-alone GPS system is not sufficient to be used for the autonomous driving system. Therefore, the curve position calibration by using in-vehicle sensors in real-time is necessary.

As an example, an autonomous driving system during right-angle cornering is shown in this paper. First, the proposed system detects the driveable region by using a LIDAR which is equipped in the front section of vehicle. Next, the right-angle corner path is found by using detected driveable region information near the right-angle corner estimated by using stand-alone GPS. Then, the desired steering angle profile to turn vehicle at the right-angle corner is determined, and this system controls the steering wheel angle with the AC servo motor.

DRIVEABLE REGION DETECTION – Fig. 5 shows the experimental location used to detect the driveable region. The vehicle ran on the road shown in Fig. 5 when the experimental vehicle was driven by a human driver, and the distance data set of the LIDAR in the vehicle-fixed coordinate system shown in Fig. 7 was measured and recorded. Fig. 8 shows the result of the data acquisition. The upper graph of Fig. 8 is the result converted into the earth-fixed coordinate system by using equation (5), and the lower graph is y-z LIDAR data in the vehicle-fixed coordinate system. The road shape can be measured by LIDAR tilted downward to the ground as shown in Fig. 8. In this paper, the driveable region is detected from the road shape configuration in the lower graph of Fig. 8. As can be noticed from the lower graph of Fig. 8, the flat part of the dot sequence which shows zero value in the z-coordinate values corresponds to the road surface. This road surface is recognized the driveable region. Fig. 9 shows the result of driveable region in the earth-fixed coordinate system. White color region shows driveable region by using a LIDAR as shown in Fig. 9. As can be noticed from the result of driveable region of Fig. 9, the region where left turn is possible can be detected by using a LIDAR.

The initialization of the steering control of left turn or right turn is determined by increasing road width x_l at left turn or right turn destination detected by using a LIDAR. When x_l becomes over 1.5m, which corresponds to the tread width of the vehicle, 1.0m, and the additional margin of 0.5m, this system generates the desired steering angle and controls the steering actuator. When x_l becomes over 1.5m as shown in Fig. 9, the target waypoint B of left turn maneuver is set on the center line of left turn as a destination point. However, y-coordinate value of B point depends on vehicle trajectory traced by the desired steering angle described below.

DESIRED STEERING ANGLE PROFILE - To improve the curve tracing performance of autonomous driving system, a feedforward steering angle term is added to the desired steering angle command value. In this system, a trapezoidal shape is used for the desired steering wheel angle as shown in Fig. 10 and the whole shape of the desired steering angle is determined according to the following conditions.

- The velocity is constant while conducting left turn or right turn maneuver
- The vehicle attitude angle difference after left turn or right turn is 90 degrees

In case that the dynamic characteristics of the vehicle and the steering actuator system are negligible, by the conditions above, the command signal profile of the steering wheel angle input can be determined depending on two parameters: the steering angle decision time "T1" and steering angle keeping time "T2" as shown Fig. 10. Here is the flow of generating the desired steering angle input using this trapezoidal steering input.

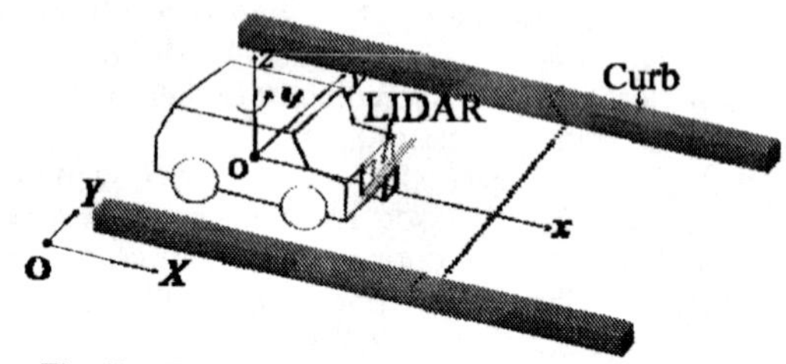

Fig. 7 Road and vehicle coordinate systems

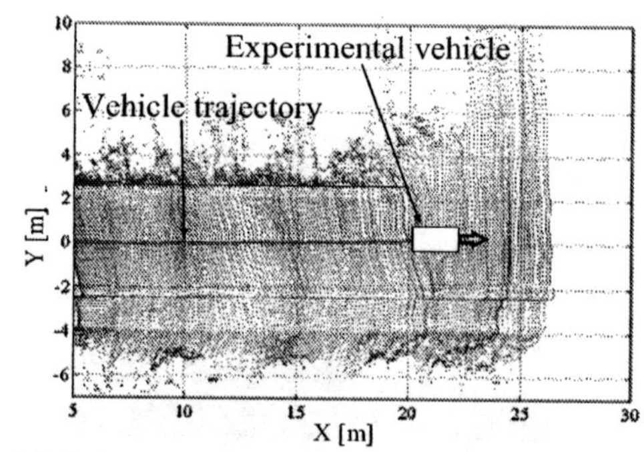

(a) Point cloud of LIDAR data in the earth-fixed coordinate system

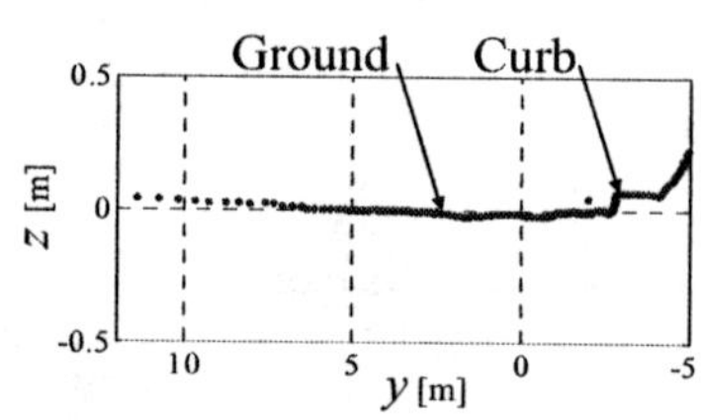

(b) LIDAR y-z data in the vehicle-fixed coordinate system

Fig. 8 Sensing of road surrounding objects with LIDAR

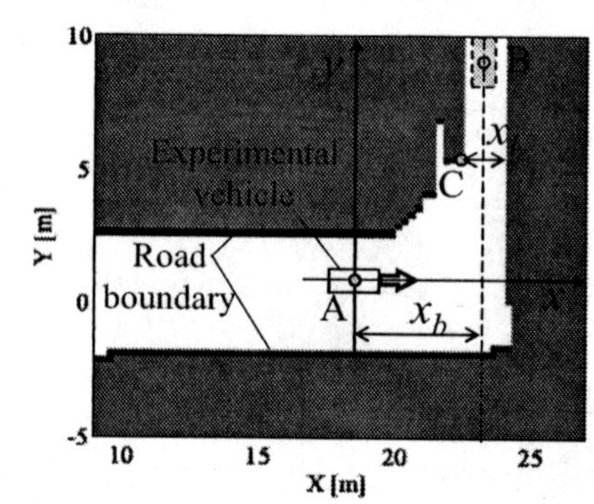

Fig. 9 Result of curve detection with LIDAR

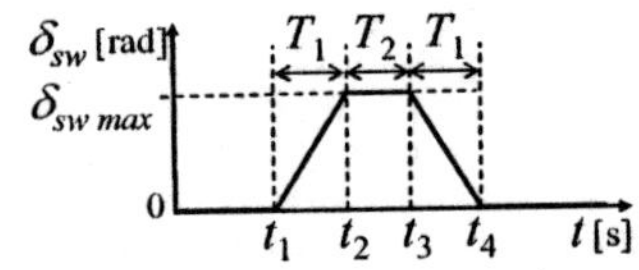

Fig. 10 Command steering angle

1. T1 and T2 counted by ΔT from zero, and a number of steering inputs are generated.
2. According to each steering angle profile, a number of corresponding predicted trajectories are calculated from the vehicle dynamics simulations.
3. Among the vehicle trajectories calculated above, candidates of left turn or right turn trajectory which is closest to the determined trajectory are selected.
4. Among the candidates of left turn or right turn trajectories, the trajectory which has the minimum yaw rate is determined as the desired left turn or right turn trajectory, and then T1 and T2 become the desired steering input parameter.

In this way, the desired value is calculated by numerical analytical approach. Therefore from the viewpoint of computational cost, it is difficult for a current general embedded controller to generate desired steering angle input to follow the left turn or right turn trajectory as soon as the system recognizes a right-angle corner. Therefore, the following procedure is used as a method to handle this problem. For steering input parameters: T1 and T2 to follow desired left turn or right turn trajectories are calculated previously offline. The parameter maps of the desired steering profile are embedded in the autonomous driving system for left turn or right turn as array data. In real-time implementation, the cornered path position can be obtained by using a LIDAR, and then the steering angle profile parameters T1 and T2 are determined by referring to the embedded array data.

RIGHT-ANGLE CORNERING SIMULATION

In order to verify the effectiveness of the proposed control algorithm for autonomous driving system, this section compares the proposed system with the previously proposed system by simulation.

SIMULATION CONDITION - Simulation condition is a left turn driving scenario on single road with the road width of 5m, and the vehicle velocity is set constantly at 10 km/h, and the road boundary can be detected without loss. In addition, the distance from the vehicle CG to the LIDAR scanning line on the road is 5.75m. The LIDAR is tilted downward at a certain degree. It is assumed that the road width of left turn at the cornering start point x_l is 1.5m, and the distance x_b, which is the distance from the vehicle CG to the target waypoint of left turn B, is 5.0 m. Therefore, in the simulation, the initialization of left turn maneuver, i.e. the waypoint A, can be determined by using the preview road shape information.

The path tracking control is switched to the lateral deviation feedback control system after the left turn. The feedback control system calculates the desired steering angle for path tracking based on the lateral deviation from the desired path at the preview point.

SIMULATION RESULT - The simulation result is shown in Fig. 11. As the left turn end point B does not exist at the road center of left turn destination, the vehicle center of gravity does not correspond with the road center after left turn as shown result of vehicle trajectory in Fig. 11. In order to solve this problem, the improvement of the curved path tracking performance can be done by feedback of vehicle lateral deviation obtained by using the LIDAR during left turn. However, in comparison with the previous system, the maximum yaw rate and the steering wheel control are decreased. As a result, the running stability of vehicle was improved in comparison with the case of using only the feedback control system. Furthermore, after the left turn maneuver is accomplished, it was confirmed that vehicle can continue following the road path by switching to the feedback control system without lane departure.

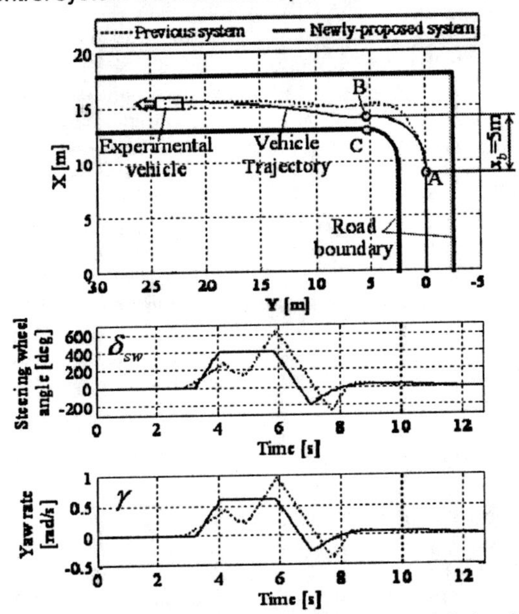

Fig. 11 Simulation result of autonomous path tracking system

RIGHT-ANGLE CORNERING EXPERIMENT

The effectiveness of the proposed desired steering angle generating system is verified by experiments using the micro electric vehicle.

EXPERIMENTAL CONDITION - The experimental course used in the experiment is shown in Fig. 12. In the curve shown in Fig. 12, the left turn autonomous driving experiment by using a LIDAR to detect the corner position was conducted. It is assumed that the proposed system can use a stand-alone GPS to know the rough vehicle position with low accuracy. The corner detection began by using the LIDAR when the vehicle approached the corner, and then the distance x_b, from the ego vehicle front end to the left turn target waypoint B, was detected. The desired steering input parameters: T1 and T2 to follow the desired left turn trajectory were determined by referring to the embedded array data. The desired steering angle calculation for straight path tracking before and after the left turn used the feedback control of the preview lateral deviation. Here, the preview path deviation was calculated by referring to the position of only left side curb, and the target path was assumed to be the parallel line to the left side road boundary with the distance of 2.50 m. Moreover, the in-wheel motor was controlled so that the vehicle velocity became constant at 10km/h.

EXPERIMENTAL RESULTS - Experimental results are shown in Fig. 13. The test drive result using the proposed system shows that the left turn can be automatically done without deviating from the lane, and the path tracking after left turn can be done without lane departure. In Fig. 13, the waypoint A indicates the left turn initialization point, and the waypoint B is the left turn destination point. As shown in Fig. 13, the distance x_b from the point A to the point B is 4.7m, calculated by using a LIDAR. From the graph of the steering wheel angle, the actual steering angle value matched well with the desired steering wheel angle determined by the distance x_b. However, the time delay of actual yaw rate response was found when compared with the desired yaw rate. As can be noticed from the vehicle trajectory shown in Fig. 13, this time delay results in the deviation at the left turn target waypoint B. The time delay of the yaw rate is caused by the dead time and nonlinearity of the steering actuator system. Therefore, further improvement of the desired steering angle generating system by considering dead time will be considered to compensate the yaw rate delay as well as the vehicle trajectory deviation.

CONCLUSIONS

This paper proposes the autonomous path tracking system including a straight path and a right-angle corner path based on the information of curb position acquired by a LIDAR. The active front steering angle is controlled to trace the desired path and determined by the feedback control of the lateral deviation from the road center and the feedforward control depending on the preview corner position. The effectiveness of the system is verified by autonomous path tracking simulation and test drives by using micro electric vehicle. During cornering, the maximum yaw rate of the proposed system is smaller than the case of using only the feedback control system, as well as the yaw oscillation is reduced.

As further improvement of the system, the desired steering angle profile considering the dead time and other nonlinearities of the steering actuator will be studied. The integration with the lateral feedback control during cornering will be designed, and the combination with the speed control in cornering will be studied.

REFERENCES

[1] Institute for Traffic Accident Research and Data Analysis (ITARDA), Analysis of single vehicle crash, No.80, 2009.

[2] Omae, M. "Development of Multi Purpose Small Electric Vehicle with Application of Automated Guiding Control System", Journal of Asia Electric Vehicle (JAEV), Vol. 2, No. 1, 2004, pp. 557-563.

[3] Kato, S., Hashimoto, N. and Tsugawa, S. "A Course Map Generation and Employment of the Map for Autonomous Driving on Precincts Roads", Trans. of JSAE, Vol. 40, No. 5, 2009, pp. 1381-1386. (in Japanese)

[4] Yoshizawa, T. Pongsathorn, R. and Nagai, M. "A Path Tracking Control System of Autonomous Vehicles in Urban Area Based on LIDAR Information", 10th International Symposium on Advanced Vehicle Control (AVEC'10), 2010, pp. 924-929.

Fig. 12 Road used for cornering experiment

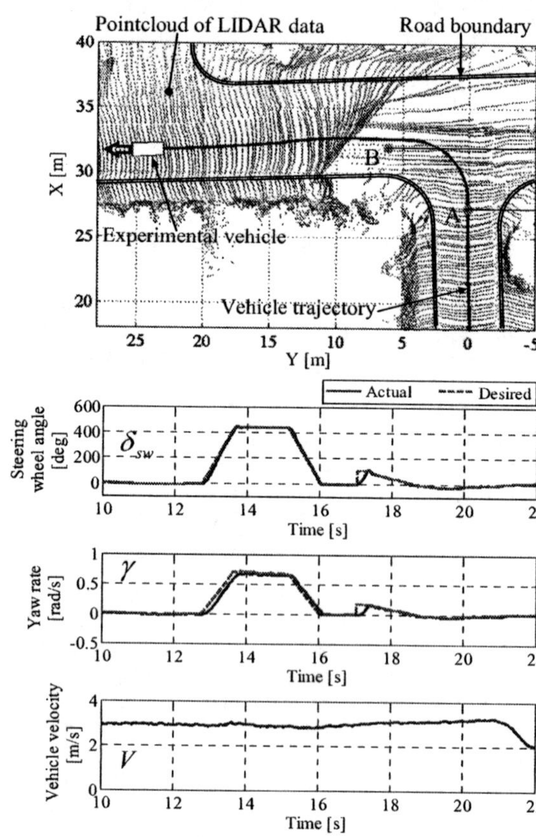

Fig. 13 Experimental result of autonomous path tracking system

SAE International

Navigation Control in an Urban Autonomous Ground Vehicle

2011-01-1037
Published
04/12/2011

Bapiraju Surampudi and Joe Steiber
Southwest Research Institute

doi:10.4271/2011-01-1037

ABSTRACT

Southwest Research Institute developed an Autonomous Ground Vehicle (AGV) capable of navigating in urban environments. The paper first gives an overview of hardware and software onboard the vehicle. The systems onboard are classified into perception, intelligence, and command and control modules to mimic a human driver. Perception deals with sensing from the world and translating it into situation awareness. This awareness is then fed into intelligence modules. Intelligence modules take inputs from the user to understand the need to navigate from its current location to another destination and, then, generate a path between them on urban, drivable surfaces using its internal urban database. Situational awareness helps intelligence to update the path in real time by avoiding any static/moving obstacles while following traffic rules. Control modules take the path command from intelligence and actuate the accelerator/brake pedal and steering to physically drive the vehicle from point A to point B. Sliding mode algorithms developed for controlling the steering are described. Performance improvement of the vehicle maneuvering curves is quantified relative to conventional PID algorithms.

INTRODUCTION

Although automation has had significant technology advances in the last few decades, machine intelligence cannot yet drive through a cluttered urban environment in a reliable manner. Even emulating limited human driver capabilities has value in applications dealing with national defense, executing hazardous operations, and assisting human drivers in case of emergency or various degrees of loss of consciousness. With the advent of drive-by-steering and drive-by-wire pedals in commercial vehicles, it is easier for autonomous failsafe systems to provide driver assistance in case of emergency. A few examples include steering to the shoulder of the road when the driver has a heart attack or otherwise impaired, informing emergency medical services, signaling driver alerts and controlling steering to avoid collisions in blind zones, and automatically applying brakes to prevent collisions with vehicles in the same lane. As a vehicle testing technology autonomy can be used to run the production vehicles on rough pavements to study accelerated aging process of chassis and suspension components. Government programs such as DARPA grand/urban challenges, European Land-Robot Trial, Student Autonomous Underwater Challenges (SAUCE) have accelerated the technology development and raised awareness of the benefits autonomous driving technology. In aerospace autopilot feature is already common place in cases of difficulties to navigate or other emergencies. Unmanned aerial vehicles are used in defense operations to undertake risky missions where loss of life could be prevented. Underwater autonomous submarines assist in research and help with hazardous operations such as the oil spill emergency in the Gulf of Mexico. It is hoped that this paper adds value to the body of excellent literature in this technology area.

The paper is organized into three sections after introduction. In section 2, the hardware and software used to build the SwRI autonomous vehicle is described. In section 3, the organization software system organization is described. Section 4 deals with detailed descriptions of implementation of sliding mode control algorithms to execute commanded paths. In the last section summary and conclusions are given.

OVERVIEW OF HARDWARE AND SOFTWARE

The various building blocks used to integrate the SwRI autonomous vehicle are shown in Figure 1. A 2006 Ford Explorer was used as the vehicle platform. High-performance blades cluster with Intel Core 2 Duo processors were used for implementing intelligence decisions. Real Time Mines Automotive Prototyping System (RTMAPS™), a software tool from INRIA in France, was used to develop the intelligence software. Dedicated Short-Range Communication (DSRC) radios were used for vehicle to infrastructure (V2I) and vehicle to vehicle (V2V) communications. An Oxford™ RT3000 INS/GPS system with high positional accuracy was used for providing feedback on latitude, longitude, and heading of the vehicle. Prosillica™ high-resolution cameras were used for vision based perception. IBEO laser scanners were used to detect obstacles in the path of the vehicle. A handicap drive-by-wire system supplied by EMC was adapted for this vehicle to drive accelerator pedal, brake, and steering. A dSPACE™ autobox™ was used to manage communications, acquire sensor signals, and send real-time actuation control signals. Mathworks™ software platform, Matlab™/Simulink™, was used to implement the control algorithms for controlling pedals and steering.

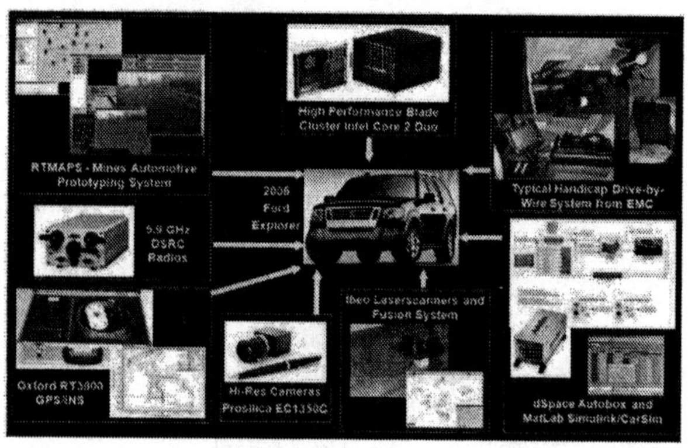

Figure 1. Building blocks of SwRI Autonomous Vehicle

The general physical locations of components are shown in Figure 2.

Figure 2. Locations of sensors, actuators and computing hardware in the vehicle

ORGANIZATION OF INTELLIGENCE

The systems organization of software modules is shown in Figure 3. On a broad basis, it is divided into perception, intelligence and CCC (Command, Communication and Control) modules. Perception deals with acquiring information from sensors, classifying, and filtering to a format easily usable by intelligence for decision making. Driver interface allows a user of the vehicle to point to a destination location on a touch screen map. The intelligence module uses the perception information and destination information to plan a safe urban path based on a database of navigation worthy roads. It also has the capability to adaptively change the path when moving or static obstacles are encountered including reflex-like behaviors. The CCC module takes perception information and path command from intelligence module and computes commands to the pedals and steering using appropriate control algorithms. It is also responsible for managing CAN traffic and failure protection.

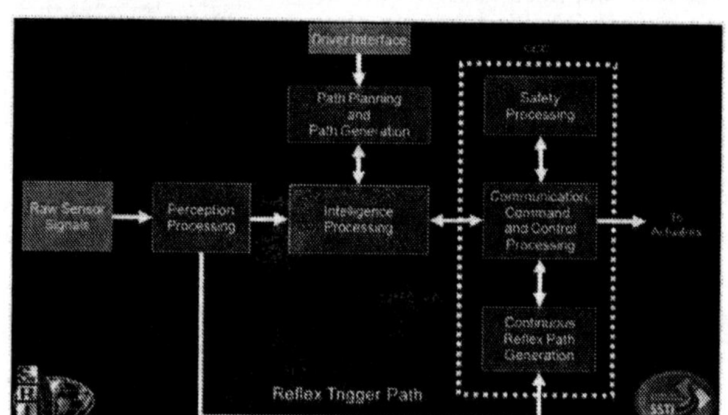

Figure 3. Autonomous System is organized into Perception, Intelligence, and CCC (Command, Communication and Control) Modules

CONTROL SYSTEM DEVELOPMENT FOR NAVIGATION ON CURVES

One of the responsibilities of the CCC module is to navigate commanded path with minimal position and heading error. This is more challenging on the curved paths. This paper will now focus on describing implementation of a sliding mode control algorithm to improve positional accuracy compared to PID controller also developed and calibrated by the authors.

GAIN SCHEDULED PID (PROPORTIONAL INTEGRAL DERIVATIVE) CONTROLLER

The PID controller developed is shown in Figure 4.

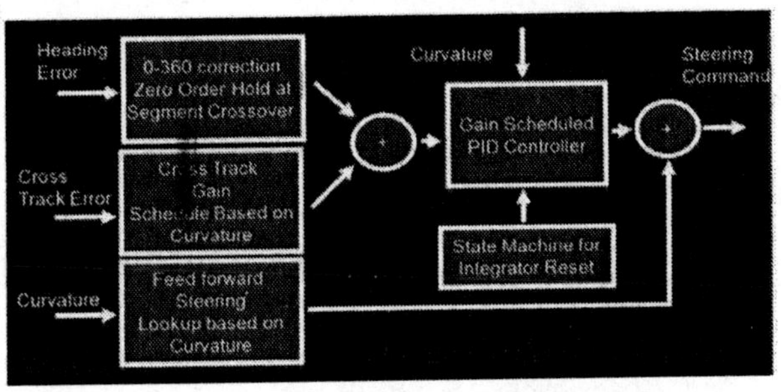

Figure 4. PID Controller for navigation path tracking

To stay on desired path at any given speed, the controller has to use error in heading angle, absolute position error (also called cross track error) and curvature to determine steering angle at any given time. The PID controller structure shown in Figure 4 reflects this input-output structure. A state machine implementation was used to manage integrators based on commanded path and state of other modules. Both feed-forward and feedback gains were scheduled based on curvature. While the PID controller performed very well despite its heuristic structure, its tracking error was excessive on curves. Also, effort needed to calibrate the controller was 3 times higher than the sliding mode controller. A curved path was navigated and the performance of the PID controller is shown in Figure 5. We can note that, during the curved portion of the path, the error can be as high as 0.6 m.

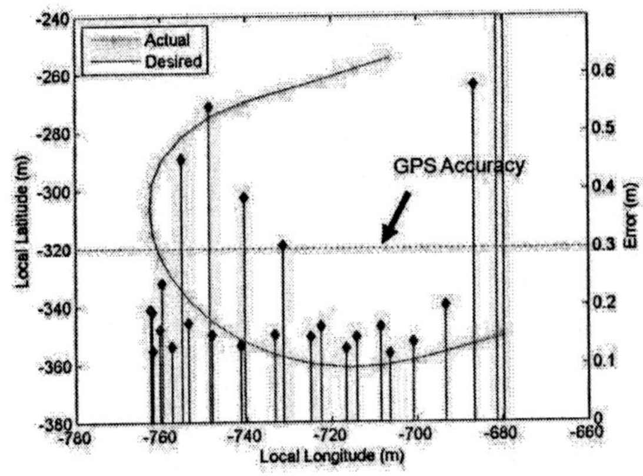

Figure 5. Tracking error of PID controller increases on curves

SLIDING MODE CONTROLLER

Sliding mode algorithm development and simulation work have been reported by the authors in [1]. The nomenclature from [1] is not repeated in this paper. The coordinates, forces, and angles used in the algorithms to follow are shown in Figure 6 and 7.

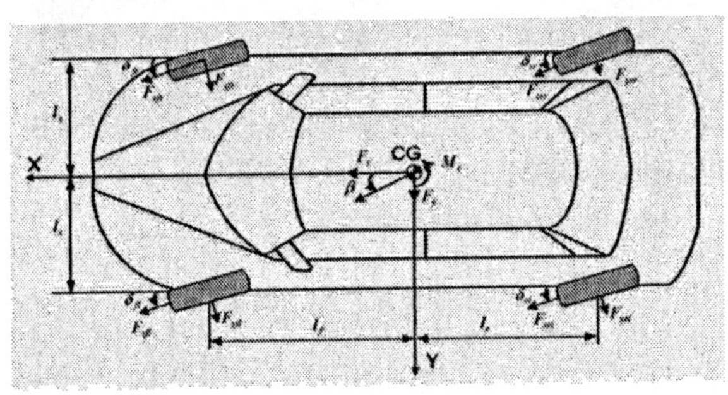

Figure 6. Global and local coordinate frames of a vehicle [From 1]

The structure of the sliding mode controller implementation is shown in Figure 8. Based on the error in latitude (LocalY), longitude (LocalX), heading, and vehicle speed, a desired yaw rate is generated. Necessary spatial-to-temporal transformation was also implemented. A sliding mode algorithm is then implemented to track this desired yaw rate rather than heading and position errors.

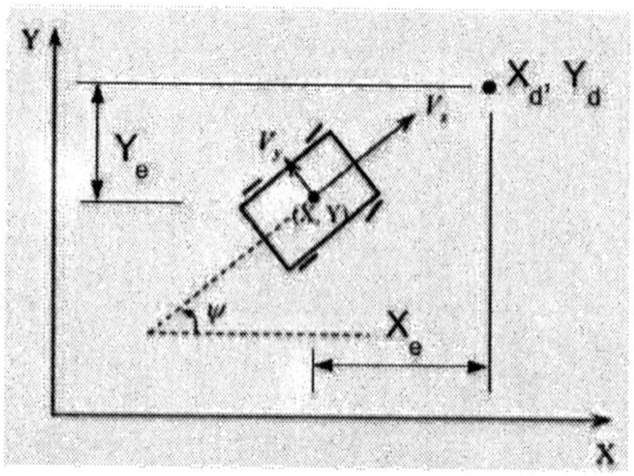

Figure 7. Error definition for X (Longitude) and Y (Latitude)

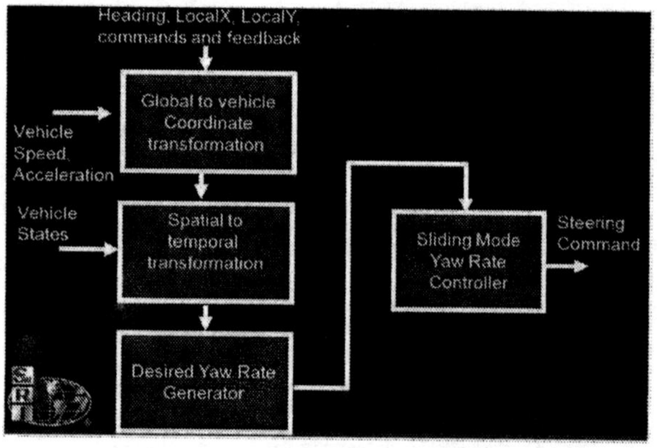

Figure 8. Implementation of sliding mode controller for path navigation

The desired yaw rate was generated based on a derivative of heading and a proportional term based on heading error as follows:

$$r_r = \dot{\psi}_{reference} + \lambda_2(\psi_{reference} - \psi) \tag{1}$$

where,

r_r is desired yaw rate

$\dot{\psi}_{reference}$ is derivative of desired heading

λ_2 is proportional term

ψ - actual heading

The sliding surface chosen for tracking the desired yaw rate is as follows:

$$S_r = (r - r_r) + \lambda_r(\psi - \psi_{ri}) \tag{2}$$

Where

λ_r is a proportional constant

Based on Lyapunov stability the steering angle was derived to be:

$$\delta_{sw} = \frac{I_z[\bar{g}(.)/I_z - \dot{r}_r + \lambda_r r - \lambda_r r_r + \eta_r \operatorname{sgn}(S_r)]}{\frac{l_f}{R_z}[\bar{F}_{zfl}K_y(\mu_{fl}) + \bar{F}_{zfr}K_y(\mu_{fr})]} \tag{3}$$

While some of the variables are shown in Figure 6 and Figure 7, the reader is referred to [1] for more detailed derivation and description.

The results of path tracking on curves for the sliding mode controller are shown in Figure 9. We can note that the error is smaller than 0.3 meters on curves for the sliding mode implementation than for PID controller.

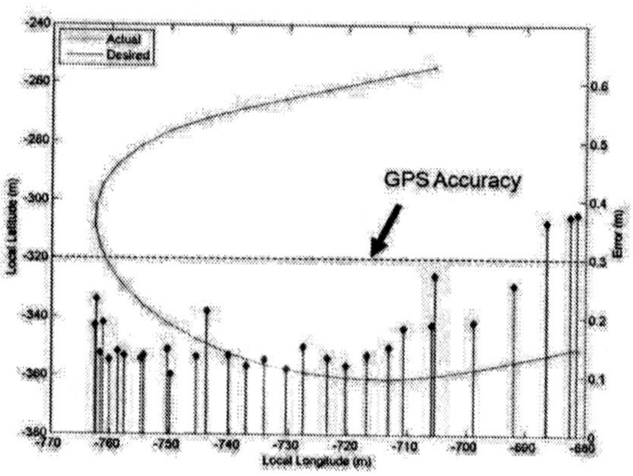

Figure 9. Sliding mode controller has low tracking error on curves as well

SUMMARY/CONCLUSIONS

An overview of the SwRI autonomous vehicle hardware and software organization is given. The challenge of tracking the commanded path on curves is shown with a conventional PID controller. Implementation of a sliding mode control algorithm, description of its structure and test results are presented. It is clearly shown that the yaw rate based sliding mode controller has superior performance compared to a PID controller. Based on the results, it is believed that the sliding mode controller implementation is better suited for applications requiring higher speeds of vehicle operation. The sliding mode controller has the potential to offer better stability control of the vehicle while requiring a manageable level of calibration.

REFERENCES

1. Wang, Junmin, Steiber, Joe and Surampudi, Bapi, "Autonomous Ground Vehicle Control System for High-Speed and Safe Operation", International Journal of Vehicle Autonomous Systems, 2008.

CONTACT INFORMATION

Please contact Bapi Surampudi at bsurampudi@swri.org or Joe Steiber at jsteiber@swri.org

ACKNOWLEDGMENTS

The authors acknowledge the contributions of SwRI autonomous vehicle team (Steve Dellenback, Ryan Lamm, Roger Lopez, Dan Pomerening, George Mc Williams, Mike Brown, Paul Avery, Steve Sprouffske, Chris Mentzer, Kris Kozak, Eric Thorn, and Josh Curtis) for the foundation work of building the vehicle. They also acknowledge Dr. Junmin Wang (now with Ohio State University) for his partnership work with Reference [1] which allowed the implementation and testing work reported in this paper.

DEFINITIONS/ABBREVIATIONS

SwRI

Southwest Research Institute

PID

Proportional, Integral and Derivative

About the Editor

After graduating from Rensselaer Polytechnic Institute with a BEE, Ronald K. Jurgen held various technical magazine editorial staff positions, including 30 years with *IEEE Spectrum*. Now retired, he is the editor of the *Automotive Electronics Handbook* and the *Digital Consumer Electronics Handbook*, and assistant editor of the *Electronics Engineers' Handbook, Fourth Edition*. He is also the editor of more than a dozen SAE International books on automotive electronics.